Eighth Edition

EXPERIMENTS IN ELECTRIC CIRCUITS

To Accompany

Thomas L. Floyd's *Principles of Electric Circuits: Conventional Current Version,* Eighth Edition,
and
Principles of Electric Circuits: Electron Flow Version, Eighth Edition

BRIAN H. STANLEY
Foothill College
Los Altos Hills, California

PEARSON

Prentice
Hall

Upper Saddle River, New Jersey
Columbus, Ohio

Acquisitions Editor: Kate Linsner
Production Editor: Rex Davidson
Design Coordinator: Diane Ernsberger
Editorial Assistant: Lara Dimmick
Cover Designer: Candace Rowley
Cover art: Getty
Production Manager: Matt Ottenweller
Senior Marketing Manager: Ben Leonard
Marketing Assistant: Les Roberts
Senior Marketing Coordinator: Liz Farrell

This book was set in Times Roman by The Special Projects Group and was printed and bound by Courier Kendallville, Inc. The cover was printed by Coral Graphic Services, Inc.

Pearson Education Ltd.
Pearson Education Singapore Pte. Ltd.
Pearson Education Canada, Ltd.
Pearson Education—Japan

Pearson Education Australia Pty. Limited
Pearson Education North Asia Ltd.
Pearson Educación de Mexico, S.A. de C.V.
Pearson Education Malaysia Pte. Ltd.

10 9 8 7 6 5 4 3 2 1

ISBN: 0-13-170180-0

Dedication

To my Mother—If I had ever been given the chance to choose which mother I would have . . . I would have chosen you.

PREFACE

This eighth edition of *Experiments in Electric Circuits* retains all 53 experiments from the seventh edition. It is an important supplement to *Principles of Electric Circuits,* by Thomas L. Floyd, and each experiment provides related reading and end-of-chapter problem set references to that text.

This laboratory manual also can be used very successfully in virtually any lab setting designed for the study of electric circuits. Many users of the manual find it compatible with a wide range of other basic circuits texts, and one only has to cross-reference titles through a book's contents or index.

Illustrations in the manual employ conventional current; however, the electron flow version of some of the same illustrations can be found in Appendix C.

FEATURES

1. Early experiments in both the dc and ac sections concentrate on instrument familiarity to help overcome a common student ailment: instrument intimidation.
2. All of the experiments relate to a particular part of the associated text and contain references to both section number and section title.
3. The ac experiments have been carefully devised to minimize instrument loading and other effects that often seem to "distort the truth" for the student.
4. The manual is designed so that the collected data and experiment questions can be detached and handed in together at the instructor's discretion, leaving the rest of the experiment intact for the student's reference.
5. The questions allow a multiple choice and/or essay approach. Each experiment contains six questions focusing on the key ideas that ought to have been learned in the execution of the experiment. The last two of the six are meant to be an inspiration for written conclusions, and the instructor may want to add to these.

NOTICE TO THE READER

The publisher and the author do not warrant or guarantee any of the products and/or equipment described herein nor has the publisher or the author made any independent analysis in connection with any of the products, equipment, or information used herein. The reader is directed to the manufacturer for any warranty or guarantee for any claim, loss, damages, costs, or expense arising out of or incurred by the reader in connection with the use or operation of the products and/or equipment.

The reader is expressly advised to adopt all safety precautions that might be indicated by the activities and experiments described herein. The reader assumes all risks in connection with such instructions.

NOTICE TO THE INSTRUCTOR

Answers to all of the questions at the end of the experiments are in the Instructor's Resource Manual (print ISBN: 0-13-170182-7, online ISBN: 0-13-188972-9) that accompanies Floyd's text *Principles of Electric Circuits*. This manual is available from Prentice Hall for instructors who teach the course.

To access supplementary materials online, instructors need to request an instructor access code. Go to **www.prenhall.com**, click the **Instructor Resource Center** link, and then click **Register Today** for an instructor access code. Within 48 hours after registering you will receive a confirming e-mail including an instructor access code. Once you have received your code, go to the site and log on for full instructions on downloading the materials you wish to use.

CONTENTS

PART II: AC EXPERIMENTS   151

APPENDICES   331

EXPERIMENT GUIDELINES

All of the experiments in this manual have been tried and proven by many students before you and should give you little trouble in normal laboratory circumstances. However, a few guidelines will help you conduct the experiments quickly and successfully.

1. Each experiment has been written so that you follow a structured logical sequence meant to lead you to a specific set of conclusions. Be sure to follow the procedural steps in the order in which they are written.
2. Read the entire experiment and research any required theory beforehand. Many times an experiment takes longer than one class period simply because a student is not well prepared.
3. Once a circuit is connected, if it appears "dead," spend a few moments checking for the obvious. Some common simple errors are
   a. Equipment: power not applied
   b. Equipment: not switched on
   c. Connecting leads: loose or open-circuit
   d. Components: incorrect values
   e. Components: faulty or burned out
   f. Circuit: inadvertent connections
   g. Instruments: not calibrated
   Generally, the problems are with the operator and not the equipment.
4. When making measurements, check for their sensibility. If they appear inconsistent with what you would expect, find out why as soon as possible. You will have a schedule of experiments to complete, and time is of the essence.
5. Don't "fiddle" with your results to make them appear exactly consistent with theoretical calculations. Rather, search for answers to why they are different. Experiments will have outcomes limited by the accuracy of your instruments and your general level of skill in the laboratory. Generally, ac experiments will not yield the highly accurate results of dc experiments. Keep this in mind.
6. Above all, follow the safety rules set out by the instructor or college, and exercise care and common sense where electrical apparatus is concerned. Even low-voltage circuits can be hazardous. Be safe rather than sorry.

DATA RECORDING AND QUESTIONS

For each experiment there is a tear-out sheet containing a record of the data collected and space for you to record the answers to six questions at the end of the experiment. When completing this sheet, you should consider the following:
1. Record all quantities in appropriate units in the spaces provided.
2. Use the space headed "Notes" for any special observations that were made during the course of the experiment. (They may help you with the questions.)
3. Any graphs that accompany the tear-out sheet should be clearly labeled on the ordinate and abscissa axes. Linear relationships should be drawn as a best-straight-line-fit to the data points in question. Curves can be drawn free-hand or with mechanical aid.
4. Questions 5 and 6 require short essay answers, and you should give a clear and concise explanation. They are meant to encourage original thought, and occasionally go slightly beyond the practical components of the experiment. When a derivation is asked for, it must be clearly shown. Write in good English prose.

REFERENCE READING

As mentioned in the Preface, this book was designed to accompany Thomas L. Floyd's *Principles of Electric Circuits: Conventional Current Version,* Eighth Edition and *Principles of Electric Circuits: Electron Flow Version,* Eighth Edition.

If you are using a different dc/ac text (one not written by Floyd), your instructor can help you match an experiment with the appropriate section of your text through its table of contents or index.

If your text uses electron flow notation, you may find that the diagrams in Appendix C are less confusing than those within the experiments. The only difference between electron flow and conventional current diagrams is the direction of the arrows; the numerical results of the experiments will be identical.

SAFETY IN THE LABORATORY

The experiments in this Laboratory Manual have been designed (except for Exp. 53) for use with low voltages. Nevertheless, one should always show the utmost care when working with any electrical circuit. A current of a few mA can be lethal when passed through the body. The following rules should be observed by anyone working in an electrical laboratory to ensure the safe completion of experiments.

1. Avoid direct contact with *any* voltage source.
2. Switch off the power prior to working with any electrical circuit, no matter how low the operating voltages and currents may be.
3. Check cords for cracks, loose plugs, and frayed ends, and *never* use a cord that appears to have any of these defects.
4. Always remove watches, jewelry, and any other conductive paraphernalia before working on any circuit even if the voltage is low. Low voltage circuits are often capable of producing high currents and this, in turn, can cause severe burns.
5. Avoid working on electrical equipment with wet hands. Wet (or even sweaty) hands provide increased electrical conduction and can therefore increase the possibility of electric shock.
6. Never defeat protective devices such as fuses and circuit breakers.
7. Never work alone in the laboratory.
8. If an accident should occur, shut off power immediately. Never touch a person whose body is in contact with the electrical supply. An injured person should be kept warm and should not be given any liquids.
9. If you notice any unsafe conditions, report them immediately to your instructor, and do not work on any defective equipment until it is fixed.

dc EXPERIMENTS

1

RESISTOR COLOR CODE
AND TOLERANCE

REFERENCE READING

Principles of Electric Circuits: Section 2–5.

RELATED PROBLEMS FROM *PRINCIPLES OF ELECTRIC CIRCUITS*

Chapter 2, Problems 21 through 27.

OBJECTIVE

To become familiar with the resistor color code and the use of an ohmmeter to measure resistance.

EQUIPMENT

Digital multimeter (DMM) and/or VOM
Resistors ($\pm 5\%$): 100 Ω   18 kΩ
             470 Ω   68 kΩ
               1 kΩ   100 kΩ
             2.2 kΩ   330 kΩ
              10 kΩ   1 MΩ
Lamp no. 47 (incandescent)

BACKGROUND

The measurement of resistance is one of the more common tasks of a technician. Even when specific resistance values are not required, the ohmmeter can be used to detect a faulty component in a circuit; it can be used more specifically to determine the correct operation of lamps, fuses, switches, and any number of other components. In this experiment you will use it to check whether a number of resistors lie within the tolerance specified by their color codes. You should also take this opportunity to become familiar with the ohmmeter portion of your DMM.

Most DMMs include an ohmmeter range, usually selectable by a switch which should be set to the ohms (Ω) position. Analog voltmeters (such as VOMs) usually have to be calibrated on each range. This involves touching the probes together and "zeroing" the meter; a down-scale $\infty\ \Omega$ calibration procedure may follow this. Check with the manual or your instructor before proceeding further. Digital meters are somewhat easier to use: one need only switch to the desired range.

It is to your advantage to have some idea of the magnitude of the resistance you are about to measure, particularly with analog ohmmeters. Ordinarily, the reading on the scale has to be multiplied by some factor (e.g., 10, 100, 10,000) to get the desired reading. Too large a range for a small-valued resistor, or too small a range for a large-valued resistor, will result in inaccurate readings.

A final word. Though no power will be connected to your resistors in this experiment, in actual circuits, the power must always be turned *off* before you bring your probes into contact with the component under test.

PROCEDURE

1. The color code on each resistor defines the *nominal* value about which the tolerance is defined. The nominal value is that value of resistance that the resistor would have if its tolerance were 0 percent. Use the color code to determine the nominal value in each case and record them (smallest value first) in Table 1-1.
2. Record the tolerance and the resulting theoretical maximum and minimum values for each resistor in turn.
3. Using the DMM, measure and record the actual value of each resistor, and check to see whether or not this value falls within the tolerance. Resistors are rarely out of tolerance; if one appears to be, it could be an error in the DMM measurement. Be sure not to touch probes with your fingers when measuring resistance, since readings will be affected by body resistance!
4. Your instructor may want you to measure the resistances with the VOM and record these for comparison with those obtained from the DMM. If so, record these values in the appropriate row of Table 1-1. Be sure that the VOM is calibrated at both ends of the scale, and be aware that, on some VOMs, this needs to be done on each range.
5. Now, by taking hold of the DMM leads (one in each hand), measure and record the value of your body resistance in Table 1-2.
6. Increase your grip on the leads and note the new value for your body resistance.
7. Apply a little moisture to the points of contact with the DMM leads, and record the new value once more.
8. If you have both analog and digital meters you might repeat the above procedure with the meter you have not yet used and compare results.
9. Measure the resistance of the incandescent lamp using an appropriate ohms range on the DMM. Write it down in the Notes section of the data page.

Name _____ Date _____

DATA FOR EXPERIMENT 1

TABLE 1-1

Nominal (coded)									
Tolerance									
Maximum									
Minimum									
Measured DMM									
Measured VOM									

TABLE 1-2

Body Resistance		
Value 1	Value 2	Value 3

NOTES

QUESTIONS FOR EXPERIMENT 1

(   )   **1.** A 2.7 kΩ, 10 percent resistor has a maximum value of
         (a) 2.57 kΩ   **(b)** 2.43 kΩ   **(c)** 2.84 kΩ   **(d)** 2.97 kΩ

(   )   **2.** The *minimum* value that the resistor in question 1 could be is
         (a) 2.57 kΩ   **(b)** 2.43 kΩ   **(c)** 2.84 kΩ   **(d)** 2.97 kΩ

3. The first three bands on a resistor are colored brown, black, and red. The resistance is measured and found to be 1050 Ω. If the resistor is guaranteed to be within its tolerance specification, what tolerance might be indicated by the fourth (tolerance) band?

(   )   **(a)** 5%   **(b)** 10%   **(c)** 20%   **(d)** any of these

4. A resistor having the colored bands blue-gray-orange-silver has a minimum value of

(   )   **(a)** 6.12 kΩ   **(b)** 68 kΩ   **(c)** 61.2 kΩ   **(d)** 64.6 kΩ

5. Why did your body resistance decrease from step 5 to step 7?

6. A resistor is required for an application that requires its value to be no less than 7.1 kΩ and no more than 7.9 kΩ. Determine the standard value and the tolerance of the resistor you could use. Explain. See Table B-2 (Appendix B) for standard values.

2

VARIABLE RESISTORS

REFERENCE READING

Principles of Electric Circuits: Section 2–5.

RELATED PROBLEMS FROM *PRINCIPLES OF ELECTRIC CIRCUITS*

Chapter 2, Problem 28.

OBJECTIVE

To investigate the properties of potentiometers. To investigate the resistance versus temperature variation of a thermistor.

EQUIPMENT

DMM or VOM
Linear 1 kΩ *or* 10 kΩ ten-turn potentiometer
Thermistor (resistance between 1 kΩ and 10 kΩ at room temperature)
Source of heat, such as low-wattage soldering iron

BACKGROUND

Often electronic circuits require the user to vary the resistance between two points. This is accomplished with a device called a *rheostat*. A more useful device is the *potentiometer*. By positioning a wiper on a carbon track, a resistor can be split into two resistance values on a continuous basis such that the sum is always equal to the total. In this experiment you will investigate the properties of potentiometers. There are two basic kinds, linear and nonlinear. In linear potentiometers, the amount of resistance between either end point and the wiper (Figure 2-1) is proportional to the wiper position and, therefore, the number of turns swept through. In the nonlinear type, the track is tapered, so equal increments in distance along the track (and therefore rotation) do not yield equal changes in resistance.

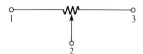

FIGURE 2-1

The second device we will look at in this experiment is the thermistor, or heat-sensitive resistor. Its resistance may increase or decrease with increasing temperature. If you are using a soldering iron, pay special attention, as always, to your activity and take safety precautions.

PROCEDURE

Part A: Potentiometer

1. Identify the end terminals and wiper terminal of the potentiometer. Number them 1, 2, and 3, with 2 being the wiper. Set the DMM (VOM) to **OHMS.**
2. Position the DMM across terminals 1 and 3. Record the measured resistance in Table 2-1 under R_{13}.
3. Position the DMM across terminals 1 and 2, then across 2 and 3. Record the respective readings in the table.
4. Add the values under R_{12} and R_{23}. Their sum should be a value close to the resistance measured in step 2.
5. Turn the shaft of the potentiometer and repeat steps 3 and 4. Record all results in Table 2-1.
6. Turn the shaft fully counterclockwise and locate the pair of terminals (1–2 or 2–3) that gives a value of resistance close to 0 Ω.
7. Give the shaft of the potentiometer one single turn clockwise, then measure and record the resistance you read across the terminals you identified in step 6.
8. Give the shaft another turn clockwise. Then once again measure and record the resistance value.
9. Repeat step 8 until the shaft will turn no farther, recording each measured value of resistance in Table 2-2. Try to visualize what is going on with respect to the circuit symbol and the actual device.

Part B: Thermistor

1. Measure and record in the Notes section of the data sheet the resistance of the thermistor at normal room temperature.
2. With the DMM still connected across the thermistor, apply heat to the body of the thermistor with the tips of your fingers and notice the change in resistance. Record the new value of resistance in the Notes section of the data sheet.
3. You can get the thermistor's resistance to change more dramatically by applying the heat from a low-wattage soldering iron or heat blower. If your instructor permits, try this, and again record the new resistance value (the resistance value may fluctuate).
4. Is the thermistor of the positive or negative temperature coefficient type?

Name _____ Date _____

DATA FOR EXPERIMENT 2

TABLE 2-1

Resistance Measured		
R_{13}	R_{12}	R_{23}

TABLE 2-2

Number of Turns	Resistance
1	
2	
3	
4	
5	
6	
7	
8	
9	
10	

NOTES

QUESTIONS FOR EXPERIMENT 2

1. Given a ten-turn linear potentiometer of value 50 kΩ, what would be the value of resistance measured between terminals 1 and 2 (in Figure 2-1)?
   (a) 50 kΩ   **(b)** 25 kΩ   **(c)** 0 kΩ

( )   **(d)** It would depend on the position of the wiper

2. As the wiper is moved from left to right in Figure 2-1, the resistance between terminals 1 and 2
   (a) increases   **(b)** decreases

( )   **(c)** increases, then decreases   **(d)** stays constant

3. As the wiper is moved from right to left in Figure 2-1, the resistance between terminals 1 and 3
   (a) increases   **(b)** decreases

( )   **(c)** increases, then decreases   **(d)** stays constant

4. If the resistance between terminals 1 and 2 of the potentiometer were 2 kΩ, and the resistance between terminals 1 and 3 were 10 kΩ, then this would be called a (an)
   (a) 10 kΩ pot   **(b)** 2 kΩ pot

( )   **(c)** 12 kΩ pot   **(d)** 8 kΩ pot

5. Detail (with a sketch) how you would connect the potentiometer in a rheostat configuration.

6. Explain in your own words the difference between a positive temperature coefficient and negative temperature coefficient thermistor. What would be the significance of a zero temperature coefficient resistor?

SWITCHES AND FUSES

REFERENCE READING

Principles of Electric Circuits: Sections 2–5, 3–20.

RELATED PROBLEMS FROM *PRINCIPLES OF ELECTRIC CIRCUITS*

Chapter 2, Problems 21 through 26.

OBJECTIVE

To examine the operation of various kinds of switches and to understand the function of fuses in electric circuits.

EQUIPMENT

Ohmmeter
Normally open and normally closed push buttons
Single-pole/single-throw and single-pole/double-throw switches
A selection of intact and "blown" fuses
Lamp no. 47 (incandescent)

BACKGROUND

Switches and fuses, together with resistors and energy sources, make up a large part of electronic circuitry. Switches are simple control mechanisms that allow current flow when closed and inhibit current flow when open. Though you will look at a few mechanical switches in this experiment, many electronic devices such as diodes, transistors, and integrated circuits (ICs) can be made to behave like switches. A number of different varieties are available, and with an ohmmeter you can easily categorize and identify the various types. To begin, we will examine the four basic kinds shown in Figure 3-1. These are the normally open and normally closed push buttons (NOPB and NCPB, respectively), and the single-pole/single-throw (SPST) and single-pole/double-throw (SPDT) switches.

11

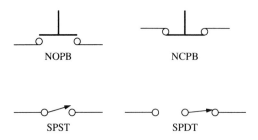

NOPB   NCPB

SPST   SPDT

FIGURE 3-1

A fuse is like a switch that automatically opens and halts the current when it exceeds a safe or normal value in a circuit. It consists of a fine metallic filament that melts when its temperature is raised to some predetermined level. In the experiment, you will use an ohmmeter to determine whether a fuse is intact or "blown."

PROCEDURE

1. Identify the four switch types and associate each one with its corresponding symbol in Figure 3-1.
2. Take the two push-button switches and use the ohmmeter to determine which is normally open and which is normally closed. What readings do you obtain for (a) a closed switch, and (b) an open switch?
3. Measure the resistance between the terminals of the single-pole/single-throw (SPST) switch for each position of the switch.
4. Identify which pairs of terminals are connected for the two positions of the single-pole/double-throw (SPDT) switch. Use your ohmmeter as in step 3.
5. Using the most sensitive scale on your ohmmeter, determine the typical resistance of an intact fuse. Determine also the resistance of a blown fuse. Record these data in Table 3-1.
6. With the power supply switched off, connect the circuit in Figure 3-2. The switch is an SPST type. Make sure that the switch is in the open (off) position. You should be able to use the DMM on **OHMS** to determine when this is the case. Make sure that you test the switch when it is *not* in the circuit.
7. Turn on the power supply and, using the panel meter (a meter on the front panel of the power supply), adjust the voltage to a value of 5 V. Measure the voltage across the switch terminals. Its value should be 5 V. (Why?)*
8. Flip the switch to its closed (on) position. The lamp should light. Measure the voltage across the switch terminals. Its value should now be 0 V. (Why?)*

---

*You may not be able to address this question unless you have already covered Ohm's Law and K.V.L.

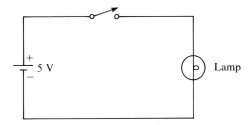

FIGURE 3-2

9. Switch off the power supply and replace the switch with the NOPB. Switch on the power and observe the lamp condition when the button is pushed and released.

10. Once again, switch off the power supply and replace the NOPB with the NCPB; repeat step 9.

Name _____ Date _____

DATA FOR EXPERIMENT 3

TABLE 3-1

	Intact Fuse	Blown Fuse
Measured Resistance (Ω)		

NOTES

QUESTIONS FOR EXPERIMENT 3

( )

1. When a normally open push button is depressed, the resistance between its terminals is approximately
(a) 0   **(b)** $\infty\ \Omega$ **(c)** neither of these

2. When a normally closed push button is depressed, the resistance between its terminals is approximately

( )

(a) 0   **(b)** $\infty\ \Omega$ **(c)** neither of these

3. A single-pole/single-throw switch has
(a) one position and two terminals
(b) two positions and two terminals

( )

(c) two positions and three terminals

4. A single-pole/double-throw switch has
(a) two positions and two terminals
(b) two positions and three terminals
(c) three positions and three terminals

( )

(d) two positions and six terminals

5. Which kind of switch would you use to facilitate the on/off action of a light switch? Make a sketch of such an application, using a simple battery resistor circuit for your example.

6. Give an example of a case when a momentary push-button switch is preferred over a latching (SPST) switch.

4

VOLTAGE AND CURRENT MEASUREMENTS IN AN ELECTRIC CIRCUIT

REFERENCE READING

Principles of Electric Circuits: Sections 2–7, 2–8.

RELATED PROBLEMS FROM *PRINCIPLES OF ELECTRIC CIRCUITS*

Chapter 2, Problems 37 through 47.

EQUIPMENT

dc power supply 0–10 V
DMM or VOM
Resistor (±5%): 1 kΩ

BACKGROUND

Information about voltage and current values in an electronic circuit is crucially important in the determination of faulty components. Measurement of voltage, in particular, is the most common activity in a technician's daily routine. Several different kinds of voltage- and/or current-measuring instruments are available. For many years, the VOM (volt-ohm-milliammeter) was most commonly seen. It is gradually being replaced by portable electronic voltmeters with digital readouts.

This experiment will familiarize you with basic *voltage* and *current* measurements *across* and *through* a resistor that is connected to a dc power supply. We shall deal with the measurement of voltage first. There are two main points to remember: the method of connecting the voltmeter, and the polarity. A voltmeter must always be connected with the probes *across* the component under test; that is, the circuit need never be broken (Figure 4-1). This is often referred to as a *parallel connection*.

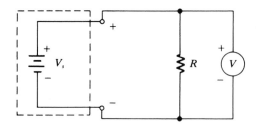

FIGURE 4-1

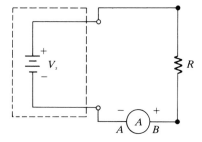

FIGURE 4-2

Knowledge of the polarity of a voltage is essential when using an instrument such as the VOM (or any analog meter, in general). In addition, its function switch must be in the correct position for an upscale reading (one in which the pointer moves to the right). Usually +dc means that the meter will read upscale if the *red* lead is *positive* and the *black* (or common) lead is *negative,* and −dc means that the meter will read upscale if the *red* lead is *negative* and the *black* lead is *positive.* DMMs do not normally have a polarity switch and merely indicate a negative voltage when inserted "backwards."

In the measuring of current, the meter must always be inserted *into* the circuit in such a manner that the current to be measured will flow *through* the meter. (This is often referred to as a *series connection.*) You *must* break the circuit to perform a current measurement (Figure 4-2). If you have not broken the circuit to insert the ammeter, then it is probably not in place correctly. Once again, for an analog meter you must be sure to insert it with the correct polarity: the most positive point should be connected to the positive terminal when the function switch is set to +dc.

A final word of warning. *Never* connect any meter—digital or analog—into a circuit as a voltmeter—i.e., in parallel—while it is set up as an ammeter (for measuring current). Such a connection will almost certainly blow a fuse, and can, on occasion, damage some instruments.

PROCEDURE

1. Turn your attention to the power supply. Locate a pair of output terminals together with an associated voltage level output control. There should also be a panel meter associated with the output terminals. This meter can normally be used to read the voltage (and sometimes the current) at the output terminals of the supply, though it may not give as accurate a reading as your DMM. Always use the most accurate instrument available to set up a voltage level.
2. Switch on the power supply, and adjust the voltage level control for minimum output.
3. Set up the meter to measure dc voltage (**dc VOLTS** function). Choose a range on which a voltage of 1 V can easily be measured.
4. Connect the meter directly across the power supply terminals, making sure that the polarity of the meter is correct. The **NEGATIVE (−)**, or **COMMON,** side of the meter should be connected to the **NEGATIVE (−)**, or **COMMON,** side of the power supply; the **POSITIVE (+)** sides of the meter and power supply should be connected together.
5. Slowly turn up the voltage level control, and watch the meter reading. It should indicate that the voltage across the output terminals is being varied.
6. Adjust the terminal voltage to a value of 1 V.

7. Temporarily remove the meter, and connect the 1 kΩ resistor across the output terminals, as shown in Figure 4-1. Resistors have no polarity and therefore it does not matter which way around you connect in the resistor.

8. Reconnect the meter across the resistor as shown, again with due care for the polarity of the meter. The voltage across the power supply terminals should not change when the resistor is connected, so the meter should continue to read 1 V.

9. Disconnect the meter, leaving the remainder of the circuit intact, and switch to a current range (**dc mAMPS**). The current range should be one on which it will be easy to measure a current of 1 mA or so. (If using a VOM, do not select the 1 mA range—the current will be in the vicinity of 1 mA, and we do not want to "peg" the meter.)

10. Break the circuit at the negative end of the resistor, as shown in Figure 4-2, and insert the instrument set to the appropriate current range. Once again, the polarity of the meter should be observed. The **NEGATIVE (−), or COM-MON,** end of the meter should be connected to the **NEGATIVE (−), or COM-MON,** side of the power supply; the **POSITIVE (+)** side of the meter should be hooked up to the resistor.

11. The meter should now indicate the current flowing in the circuit; this should be in the neighborhood of 1 mA.

12. Should the deflection on an analog ammeter be too small to read, switch to the next smaller range. If the reading on the digital meter is not using the leading (leftmost) digits in the display, then switch to a smaller range.

13. Record the measured current in Table 4-1 under the 1 V column heading.

14. Now remove the instrument, and reconnect the bottom end of the resistor to the power supply. Reset the instrument to **dc VOLTS,** and increase the power supply voltage until you have 2 V across the resistor.

15. Disconnect the meter, set it for current, and insert it into the circuit to measure the current. Measure and record the current.

16. For each voltage in Table 4-1, measure and record the current through the 1 kΩ resistor.

17. Finally, if you have been using a DMM, count the number of digits in the instrument's display. Then, by referring to the operator's manual or consulting your instructor, determine how many of these digits are *full digits* (i.e., can go from 0 through 9) and whether or not the first digit is only a *half digit* (can go from 0 to 1).

18. In the Notes section of the data sheet, record the resolution (the smallest change in voltage that the meter can display) on each of the DMM's voltage and current scales you used in this experiment.

Name _____ Date _____

DATA FOR EXPERIMENT 4

TABLE 4-1

V_s (V)	1	2	3	4	5	6	7	8	9	10
I (mA)										

NOTES

QUESTIONS FOR EXPERIMENT 4

1. For an ammeter to correctly read the current flowing through a resistor, it must be connected
   (a) in series with the resistor
   (b) in parallel with the resistor

   ( )   (c) either **a** or **b**   (d) neither **a** nor **b**

2. For a voltmeter to correctly read the voltage across a resistor, it must be connected
   (a) in series with the resistor
   (b) in parallel with the resistor

   ( )   (c) either **a** or **b**   (d) neither **a** nor **b**

3. In the circuits of Figures 4-1 and 4-2, an increase in the power supply terminal voltage always caused
   (a) an increase in the current through the resistor
   (b) a decrease in the current through the resistor

   ( )   (c) no change in the current through the resistor

4. In connection with instruments that you might have used in this experiment, the instrument with the greatest accuracy (hint: see *Principles of Electric Circuits,* Section 2–7, or the operator's manual for the instruments) is
   (a) the DMM   (b) the VOM

   ( )   (c) both are generally equally accurate

5. Define the resolution of a DMM and give an example.

6. Show, with the aid of a diagram, another place in Figure 4-2 where the ammeter can be inserted to measure the resistor current.

OHM'S LAW

REFERENCE READING

Principles of Electric Circuits: Sections 3–1 through 3–6.

RELATED PROBLEMS FROM *PRINCIPLES OF ELECTRIC CIRCUITS*

Chapter 3, Problems 1 through 10 and 13 through 33.

OBJECTIVE

To prove that current I and voltage V are linearly proportional in a dc circuit. To show that the proportionality constant is equal to the resistance R of the circuit.

EQUIPMENT

dc power supply 0–10 V
DMM or VOM
Resistors ($\pm 5\%$): 1 kΩ
                    2 kΩ

BACKGROUND

Ohm's Law is the basis of many electrical circuit calculations and is one of the most important theories you will learn: $V = IR$. The purpose of this experiment is to verify Ohm's Law, which, in words, simply says that the current through a resistor is proportional to the voltage across it. The way in which we accomplish this is to measure the voltage across and current through a known resistor for several different pairs of values. We can then plot the data on a graph, and if the relationship is truly linear, it should yield a straight line.

  When graphing data such as those obtained in this experiment, either the x or y axis can be chosen to display the voltage or current values. When the y axis is chosen as the voltage axis, and the x axis as the current axis, we say that we are plotting V versus I. The slope of the line $\Delta V/\Delta I$ (= rise/run) should be equal to the resistance R of the resistor. If, on the other hand, current is plotted on the y axis, and voltage along the x axis, then the slope of the line $\Delta I/\Delta V$ (= rise/run) is equal to the conductance G of the resistor. In this experiment, you will first plot V in volts versus I in mA, and therefore the slope will be the conductance of R (see PEC-3, section 3–5). Then, if you wish, additional graph paper is provided to plot I versus V, so that you can verify that the slope of these data is, in fact, the resistance. When plotting a straight line on a graph such as this, it is important to remember that the object here is to prove Ohm's Law, and therefore it is imperative that you draw the best straight line that you can through the data points. *Do not join each pair of data points individually by straight lines as some students are in the habit of doing.*

Finally, this experiment goes much faster with two meters, so that one is assigned the ammeter and the other the voltmeter. This saves switching meters to get the pairs of data points.

PROCEDURE

1. With the DMM or **OHMS,** measure the resistance of both the 1 kΩ and 2 kΩ resistors, and record these measured values in Tables 5-1 and 5-2.
2. With the power supply initially off, connect the circuit in Figure 5-1 with R = 1 kΩ.
3. If two meters are available, connect one in series as an ammeter and the other in parallel as a voltmeter, as shown in Figure 5-2. Otherwise, in the following steps, set up and measure the voltage first, and then insert the ammeter to determine the current.
4. Switch on the power supply. Beginning at 0 V, increase the voltage in 1 V steps and measure and record the resulting current in Table 5-1.
5. Repeat this procedure with the 2 kΩ resistor in place of the 1 kΩ resistor. Record these data in Table 5-2.
6. On the same scales and axes, plot graphs of V versus I for each resistor. (Assign V to the y axis and I to the x axis.) Draw the best straight lines possible through each set of data. You should have two straight lines of differing slopes. Use all of the graph paper.
7. Determine the slope of each line by constructing a right-angled triangle under the line and taking the ratio $\Delta V/\Delta I$. This ratio should equal the resistance R of the resistor in question. Record these values in Table 5-3 for each set of data.
8. Because R = $1/G$, you can determine the conductance of each resistor by taking the reciprocal of the resistance. Do this for each set of data, and record this as G in the appropriate area of Table 5-3.
9. If your instructor wishes, you can use the second sheet of graph paper to plot I versus V. The slopes of each of these lines should compare well with the values you obtained for G in step 8.

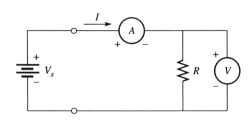

24   **FIGURE 5-1**   **FIGURE 5-2**

DATA FOR EXPERIMENT 5

TABLE 5-1

Nominal Resistance $R = 1\ k\Omega$		V_s (V)	0	1	2	3	4	5	6	7	8	9
R_{meas}		I_m (mA)										

TABLE 5-2

Nominal Resistance $R = 2\ k\Omega$		V_s (V)	0	1	2	3	4	5	6	7	8	9
R_{meas}		I_m (mA)										

TABLE 5-3

	Slope (R) (rise/run)	$G\ (1/R)$
Table 5-1		
Table 5-2		

NOTES

QUESTIONS FOR EXPERIMENT 5

(   )

1. Consider two I versus V curves on the same scales as in this experiment. Label them A and B. If the slope of A is greater than that of B, then
    (a) $R_A > R_B$ **(b)** $G_B > G_A$ **(c)** $G_B < G_A$ **(d)** $R_A = R_B$

(   )

2. If V were plotted on the vertical axis and I on the horizontal axis, then the slope would represent
    (a) R **(b)** G **(c)** neither R nor G

3. If a resistor R_B has twice the resistance of a resistor R_A, then for the same voltage, the current through R_A will be
    (a) one-half that in R_B **(b)** twice that in R_B

(   )
    (c) 4 times that in R_B **(d)** one-quarter that in R_B

4. If a resistor R_A has one-half the conductance of a resistor R_B, then for the same current, the voltage across R_A will be
    (a) twice that across R_B **(b)** one-half that across R_B

(   )
    (c) 4 times that across R_B **(d)** one-quarter of that across R_B

5. Construct a vertical line on your I versus V graph at the point $V = 5$ V. This line will intersect each of the sloping lines at two separate points. Explain the significance of these intersections.

6. Construct a horizontal line on your I versus V graph at the point $I = 1$ mA. This line will intersect each of the sloping lines at two separate points. Explain the significance of each of these intersections.

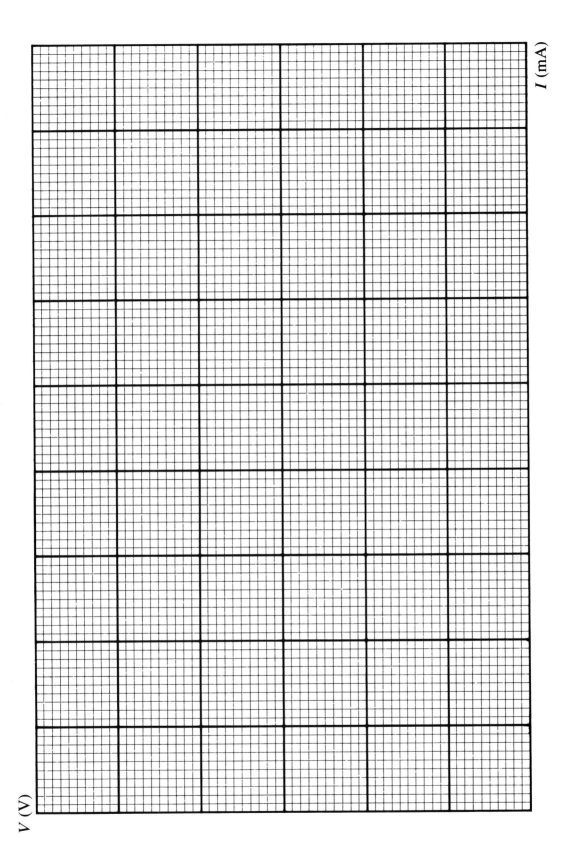

I (mA)

V (V)

27

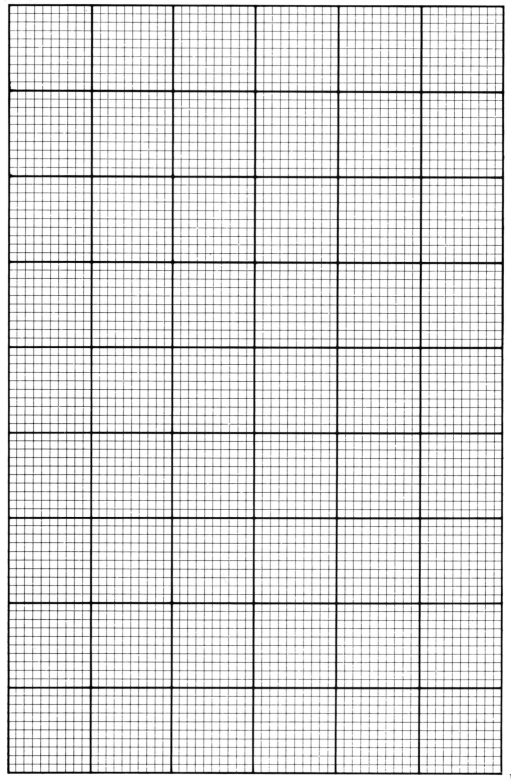

V (V)

POWER IN dc CIRCUITS

REFERENCE READING

Principles of Electric Circuits: Sections 4–1 through 4–5.

RELATED PROBLEMS FROM *PRINCIPLES OF ELECTRIC CIRCUITS*

Chapter 4, Problems 17 through 22.

OBJECTIVE

To demonstrate the three power formulae in a dc circuit.

EQUIPMENT

dc power supply 0–10 V
DMM or VOM
Resistors ($\pm 5\%$): 1 kΩ
                      2 kΩ

BACKGROUND

Electronic devices and circuits require energy to operate. Power is a measure (in watts) of the energy (in joules) consumed by a given device in one second. For a resistor, three equations will yield the power dissipated: $P = IV$, $P = V^2/R$, and $P = I^2R$. In this experiment, you will verify these formulae and plot graphs of the power versus the current, and then the power versus the voltage. The resulting curves are parabolas, and the equations of the curves are called quadratic. By plotting two curves for both the 1 kΩ and 2 kΩ resistors on the same scales and axes, you will be able to see which resistance consumes more power for a given current (or voltage).

PROCEDURE

1. Use an ohmmeter to measure the resistance of the 1 kΩ resistor. Record this measured value in Table 6-1.
2. Connect the circuit in Figure 6-1 and measure the current for each value of V_s in Table 6-1. Record these measured currents in Table 6-1.
3. Use the voltage and current data together with the measured value of R to complete the rest of Table 6-1 for the power P. All power quantities should agree.
4. Repeat steps 1 through 3 with the 2 kΩ resistor, and record all data in Table 6-2.
5. Plot graphs of P versus I for both the 1 kΩ and 2 kΩ resistors. Use the same scales and axes so that you can compare the curves. Repeat this procedure for P versus V on the second sheet of graph paper provided.

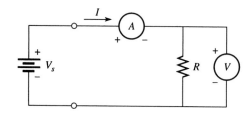

FIGURE 6-1

Name _____ Date _____

DATA FOR EXPERIMENT 6

TABLE 6-1

V_s (V)	0	1	2	3	4	5	6	7	8	9
I (mA)										
$P (= IV)$										
$P (= I^2R)$										
$P (= V^2/R)$										
	R_{meas}									

TABLE 6-2

V_s (V)	0	1	2	3	4	5	6	7	8	9
I (mA)										
$P (= IV)$										
$P (= I^2R)$										
$P (= V^2/R)$										
	R_{meas}									

NOTES

QUESTIONS FOR EXPERIMENT 6

1. In a circuit such as that in Figure 6-1, if the voltage is main-tained constant and R decreases, the power in R

(   )   **(a)** decreases   **(b)** increases   **(c)** stays the same

2. If the voltage across a resistance is doubled, the power increases by a factor of

(   )   **(a)** 2   **(b)** 4   **(c)** 1   **(d)** none of these

3. For a constant voltage, power is directly proportional to resistance.

(   )   **(a)** True   **(b)** False

4. For a P versus I curve as was plotted in this experiment, the slope is equal to the

(   )   **(a)** resistance   **(b)** conductance   **(c)** neither **a** nor **b**

5. For the same *current* values, which resistor, the 1 kΩ or the 2 kΩ, dissipates the most power? Explain how this information can be gained from your graphs of P versus I.

6. For the same *voltage* values, which resistor, the 1 kΩ or the 2 kΩ, dissipates the most power? Explain how this information can be gained from your graphs of P versus V.

P (mW)

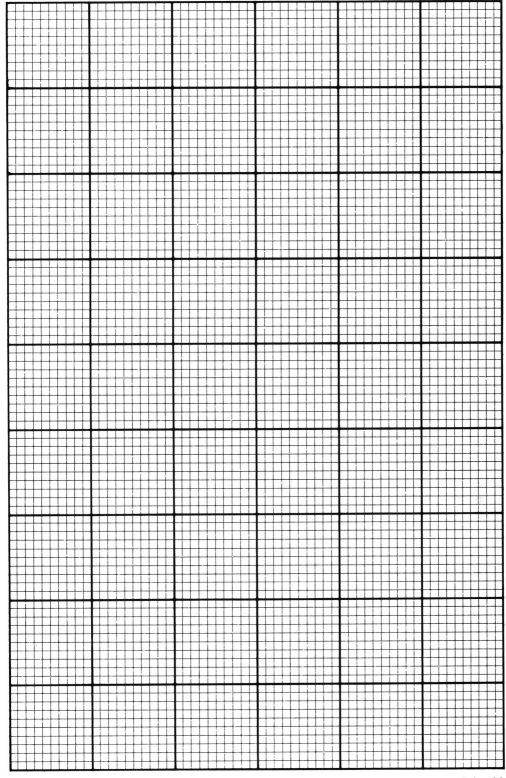

I (mA)

P (mW)

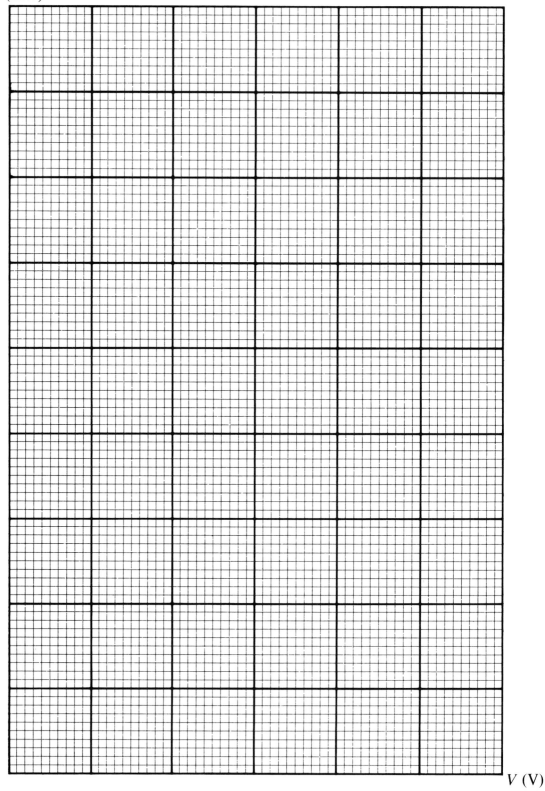

V (V)

7

RESISTOR POWER RATINGS

REFERENCE READING

Principles of Electric Circuits: Sections 4–1 through 4–5.

RELATED PROBLEMS FROM *PRINCIPLES OF ELECTRIC CIRCUITS*

Chapter 4, Problems 26 and 27.

OBJECTIVE

To determine experimentally the maximum voltage that may be applied to a resistor of given power rating without exceeding this power rating. To become familiar with the way a resistor feels to the touch when dissipating its rated power.

EQUIPMENT

dc power supply 0–10 V
DMM or VOM
Resistors (¼ W, ± 5%): 100 Ω
                            200 Ω
     (½ W, ± 5%): 100 Ω
                            200 Ω

Single-pole/double-throw switch

BACKGROUND

In experiment 6 you verified the dc power formulae. In this experiment you will determine how much power a resistor can take. The answer has to do with heat. The power in a resistor is converted to heat; the resistor is therefore warm to the touch. If the resistor should become too hot it may "fall out" of tolerance; worse still, it could ignite and cause a fire. It is important, therefore, to be aware of resistor power ratings. The power rating of a resistor is the value (in watts) that must not be exceeded if the resistor is to remain within the manufacturer's specification of ohmic value and tolerance. The power rating is a function of several variables, one of which is physical size. For example, carbon composition resistors appear in ¼ W, ½ W, 1 W, etc., sizes.

The power rating actually determines the maximum voltage and current that the resistor can safely withstand. For example, if we call the power rating P_R, then we can determine the maximum safe voltage using the dc power formula, $P = V^2/R$, and solving for V_{max}:

$$P_R = V_{max}^2/R$$

so

$$V_{max} = \sqrt{P_R R}$$

where P_R is the power rating of the resistor.

Similarly, using $P_R = I_{max}^2 R$, we can solve for the maximum safe current I_{max}:

$$P_R = I_{max}^2 R$$

so

$$I_{max} = \sqrt{P_R/R}$$

As the power rating increases, the resistor is typically larger in volume and therefore surface area. In this experiment you will become familiar with the surface temperature of various resistors as you approach their power ratings. If your instructor allows you to exceed the ratings as in the experiment, take care when touching these "hot" resistors—the heat could be unpleasant.

PROCEDURE

1. Calculate the maximum safe voltage that can be applied across your 100 Ω (¼ W) resistor without exceeding its power rating. Record this in Table 7-1.
2. Calculate the maximum safe current to which the maximum safe voltage corresponds. Record this in Table 7-1.
3. Repeat these calculations for the remaining 100 Ω resistor. Then make similar calculations for the two 200 Ω resistors. Record all calculated data in Table 7-1.
4. Connect the circuit of Figure 7-1 with $R_1 = 100$ Ω, ¼ W and $R_2 = 100$ Ω, ½ W.
5. Use the switch to connect the ¼ W resistor to the power supply voltage. Increase the voltage in 1 V increments until you reach the maximum permissible value you calculated in step 1. Pause for a few seconds at each step and touch the resistor with your thumb and forefinger. Note in particular how hot the resistor feels when the voltage is the maximum value.

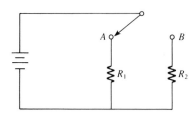

FIGURE 7-1

6. Use the switch to connect the ½ W resistor to the power supply voltage. Repeat step 5, once again feeling the temperature of the resistor with your finger.

7. With the voltage at the maximum safe value for the ¼ W resistor, compare the relative temperatures of the two 100 Ω resistors. Do this by alternately moving the switch from *A* to *B*. Be sure you allow enough time for the resistors to warm up each time. Which resistor feels warmer to the touch and why?

8. Repeat steps 4 through 7 with the 200 Ω resistors.

9. With the voltage set to the maximum safe value for the 200 Ω, ¼ W resistor, compare the relative temperatures of the two resistors as you did in step 7. Which one feels warmer and why?

NOTE: Do the following only under supervision of your instructor; the excessive heat dissipated by a resistor can cause fire or injury.

10. If your instructor allows, adjust the voltage across the ¼ W, 100 Ω resistor to about 7 V. Approximately what power does it dissipate? *Carefully* touch the resistor. Why does it feel so hot?

11. If your instructor allows, adjust the voltage across the ¼ W, 100 Ω resistor to twice its rated maximum (you calculated this in step 1). You should not touch the resistor. It will be very hot. How much power is it dissipating?

Optional Steps: For reasons of safety, we advise the instructor to do a class demonstration of these last few steps.

WARNING: There is some risk associated with this part of the experiment. The resistor may catch fire and emit fumes. However, recognition of this characteristic "aroma" might be a valuable experience in the future. It is not uncommon for resistors to overheat due to some underlying fault in a circuit. Be sure that the resistor is physically well away from the circuit protoboard, and that you keep your hands and face at a safe distance from the burning resistor. *Proceed with great caution heretofore.*

12. Increase the voltage across the resistor until you see smoke emitting from the resistor. The smoke will probably be accompanied by a characteristic smell.

13. Maintain the voltage across the resistor for a few minutes or until the smoke stops. (You can increase the voltage still further to see the resistor "ignite.")

14. Remove the resistor and allow some time for it to cool. Measure its resistance with the meter set to Ohms. How has the resistor's resistance changed?

15. The resistor has probably been irreversibly damaged and should be discarded.

DATA FOR EXPERIMENT 7

TABLE 7-1

Power Rating	Ohmic Value	Maximum Safe Voltage	Maximum Safe Current
¼ W	100 Ω		
½ W	100 Ω		
¼ W	200 Ω		
½ W	200 Ω		

NOTES

QUESTIONS FOR EXPERIMENT 7

1. The maximum safe voltage that may be safely applied to a 1000 Ω, ¼ W resistor is

( )   **(a)** 5 V   **(b)** 50 V   **(c)** 15.8 V   **(d)** 250 V

2. The maximum safe current that may safely flow through a 2000 Ω, ½ W resistor is

( )   **(a)** 0.25 mA   **(b)** 15.8 mA   **(c)** 0.354 mA   **(d)** 70.7 mA

3. You have several resistors of the same power rating but different ohmic values. Which statement is correct?

   (a) The smallest-valued resistor can safely dissipate the most power.

   (b) The largest-valued resistor can safely dissipate the most power.

   (c) The smallest-valued resistor can safely withstand the most voltage.

( )   **(d)** The smallest-valued resistor can safely withstand the most current.

4. When *tolerance* is taken into account, the largest voltage that can safely be applied to a 100 Ω, ±20%, ¼ W resistor is

( )   **(a)** 5.00 V   **(b)** 4.47 V   **(c)** 5.48 V   **(d)** 6.00 V

5. Explain in your own words why a ¼ W resistor and a ½ W resistor have different surface temperatures even when dissipating the same power.

6. If the ambient (room) temperature in which a resistor resides is lower than "normal," do you think that the resistor is able to dissipate more or less power without overheating? Give reasons for your answer.

8

SERIES RESISTIVE CIRCUITS

REFERENCE READING

Principles of Electric Circuits: Sections 5–1 through 5–4.

RELATED PROBLEMS FROM *PRINCIPLES OF ELECTRIC CIRCUITS*

Chapter 5, Problems 1 through 19.

OBJECTIVE

To verify that the total resistance in a series-connected circuit is the sum of the individual resistances.

EQUIPMENT

dc power supply 0–10 V
DMM or VOM
Resistors (±5%): 1 kΩ   3.9 kΩ
                  2 kΩ   5.1 kΩ
                  3 kΩ

BACKGROUND

When resistors are connected in series, the current that will flow is calculated from a knowledge of the total resistance. The total resistance in this case is simply the sum of the individual resistors. By knowing the total resistance, the voltage required for a desired current or the current resulting from an applied voltage can easily be calculated.

One of the purposes of this experiment is to provide you with still more experience in using your DMM. You will use it on the ohms, volts, and milliamps functions in verifying the Series Resistor Rule.

PROCEDURE

1. Calculate the upper and lower limits of the resistance for each resistor and record these in Table 8-1.
2. Use the ohmmeter to measure the *actual* values of each resistor and record these in the row labeled R_{meas}.
3. Complete Table 8-1 by calculating the upper and lower limits to R_T ($= R_1 + R_2 + R_3 + R_4 + R_5$) and then measuring the actual value of R_T by series-connecting the resistors. Does the measured value equal the sum of the individual measured values? If not, why not?
4. Using a value for V_s of 10 V, calculate the nominal, minimum, and maximum values for I in the circuit of Figure 8-1 and record them in Table 8-2.
5. Connect the circuit in Figure 8-1 and adjust V_s to a value of 10 V. Measure the current I and record it under I_{meas} in Table 8-2. (Does it appear consistent with R_{meas}?)
6. Using a value for I of 0.5 mA, calculate the nominal, minimum, and maximum values for V_s in order to sustain this current through the series combination. Record these in Table 8-3.
7. Connect the circuit and adjust V_s until I = 0.5 mA. Measure the voltage and record it in Table 8-3. (Does it appear consistent with R_{meas}?)

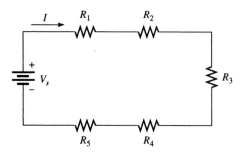

FIGURE 8-1

DATA FOR EXPERIMENT 8

TABLE 8-1

R	1 kΩ	2 kΩ	3 kΩ	3.9 kΩ	5.1 kΩ	$R_T = 15$ kΩ
R_{min}						
R_{max}						
R_{meas}						

TABLE 8-2

	I_{nom} (mA)	
$V_s = 10$ V	I_{min} (mA)	
	I_{max} (mA)	
	I_{meas} (mA)	

TABLE 8-3

	V_{nom} (V)	
$I = 0.5$ mA	V_{min} (V)	
	V_{max} (V)	
	V_{meas} (V)	

NOTES

QUESTIONS FOR EXPERIMENT 8

1. If the measured value of R_T did not equal the sum of the individually measured resistances, you might conclude that
    (a) the resistors had changed in value
    (b) the resistors were out of tolerance
    (c) the ohmmeter measurements are subject to some error

( )   (d) all of the above

2. Three resistors of value 1 kΩ, 2 kΩ, and 3 kΩ, each having a tolerance of ±5 percent, are series-connected across a 10 V source. The smallest current you would expect to flow in the circuit is
    (a) 1.59 mA   (b) 1.67 mA   (c) 1.75 mA

( )   (d) none of these

3. The current through each resistor in a series-connected circuit such as that in Figure 8-1 is the same.

( )   (a) True   (b) False

4. If the resistor tolerances in this experiment had been ±10 percent, the maximum current through the series combination with a 10 V source would have been

( )   (a) 61 mA   (b) 606 μA   (c) 741 μA   (d) 667 mA

5. If any resistor in a series circuit is reduced in value, the total current will increase. Explain why this is so.

6. If two resistors in a series circuit suffer a change in resistance such that one increases by 10 percent and the other decreases by 10 percent, explain why the current will not necessarily stay constant. Under what conditions would the current undergo no change?

VOLTAGE SOURCES IN SERIES

REFERENCE READING

Principles of Electric Circuits: Section 5–5.

RELATED PROBLEMS FROM *PRINCIPLES OF ELECTRIC CIRCUITS*

Chapter 5, Problems 27, 28, 29.

OBJECTIVE

To show that cell voltages add when in series-aiding configurations. To show that cell voltages subtract when in series-opposing configurations.

EQUIPMENT

DMM or VOM
1½ V dry cells (four)

BACKGROUND

Voltage sources of various values can be constructed using simple dry cells that can be connected in both series and parallel, as well as in aiding or opposing directions. A series connection of cells will generally produce a greater or lesser value of voltage than a single cell, depending on the polarity of the connection. A parallel connection of cells will generally provide a greater capacity for current than a single cell.

This experiment investigates a variety of interconnecting methods using four 1½ V dry cells. Follow the procedural steps carefully, and determine that you have the polarity correct and consistent with the appropriate figure before you record any voltages.

PROCEDURE

1. Identify and label your cells V_1, V_2, V_3, and V_4. Measure the terminal voltages of each of the cells and record them in Table 9-1.
2. Now connect the cells in each of the different series combinations shown in Figure 9-1. For each combination, calculate and record the expected value for the terminal voltage V_t. Then measure and record the terminal voltages in Table 9-2.

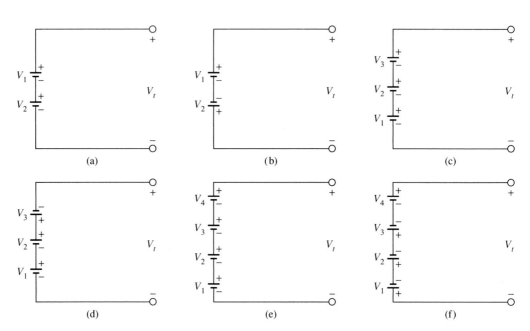

FIGURE 9-1

Name _____ Date _____

DATA FOR EXPERIMENT 9

TABLE 9-1

	Cell V_1	Cell V_2	Cell V_3	Cell V_4
Terminal Voltage V_t				

TABLE 9-2

Configuration	V_t	
	Calculated	Measured
(a)		
(b)		
(c)		
(d)		
(e)		
(f)		

NOTES

QUESTIONS FOR EXPERIMENT 9

( )

1. Referring to Figure 9-1a, if V_1 and V_2 are exactly the same value, the terminal voltage would be equal to
   (a) V_1 **(b)** $2V_1$ **(c)** $2V_2$ **(d)** 0 **(e)** **b** and **c**

2. Referring to Figure 9-1, when cells are connected in series the terminal voltage is always greater than that of a single cell.

( )

   (a) True **(b)** False

3. Referring to Figure 9-1f, if V_1 = 1.54 V, V_2 = 1.56 V, V_3 = 1.45 V, and V_4 = 1.40 V, then V_t would be

( )

   (a) -3.15 V **(b)** -2.87 V **(c)** 5.95 V **(d)** 3.15 V

4. With values as in question 3, the terminal voltage in Figure 9-1d would be

( )

   (a) 3.10 V **(b)** 4.55 V **(c)** 1.65 V **(d)** 1.45 V

5. Show how a battery with a terminal voltage of 12 V can be made up of 1½ V cells. Make a sketch of the configuration.

6. Show with the aid of a sketch how, by connecting two more cells in series with those of Figure 9-1f, the voltage V_t can be made equal to zero. Assume each cell is 1.5 V.

KIRCHHOFF'S VOLTAGE LAW

REFERENCE READING

Principles of Electric Circuits: Section 5–6.

RELATED PROBLEMS FROM *PRINCIPLES OF ELECTRIC CIRCUITS*

Chapter 5, Problems 30 through 36.

OBJECTIVE

To verify Kirchhoff's Voltage Law for dc circuits.

EQUIPMENT

dc power supply 0–10 V
DMM or VOM
Resistors ($\pm 5\%$): 1.2 kΩ   4.7 kΩ
               1.8 kΩ   5.6 kΩ
               2.4 kΩ   6.8 kΩ
               3.3 kΩ   8.2 kΩ

BACKGROUND

Kirchhoff's Voltage Law states that the algebraic sum of voltages around a closed path is equal to zero. With regard to Figure 10-1, this means that

$$V_s - V_1 - V_2 - V_3 - V_4 = 0$$
$$\text{or } V_s = V_1 + V_2 + V_3 + V_4$$

By connecting such a circuit and measuring the voltages, it should be possible to verify this relationship. In performing an experiment of this nature, you should remember that each measurement is subject to some error, and when you sum such measurements, the errors may add and produce what may appear to be an incon-

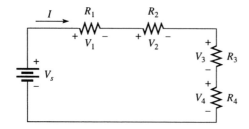

FIGURE 10-1

sistency. The important thing to remember is that, if the instruments are perfect, we should obtain a sum of voltage drops *exactly* equal to the source voltage.

PROCEDURE

1. Select four resistors from the eight available, measure their individual resistances, and record in Table 10-1.
2. Connect the resistors in series, as shown in Figure 10-1.
3. Using the measured values of resistance, calculate and record the total resistance R_T and the current I that would result from a source voltage of 10 V.
4. Use the calculated current and measured resistance values to predict the voltage drops across each of the resistors. Record these in the Calculated Voltages row in Table 10-1.
5. Add the calculated values, and record this under the V_{sum} heading in the Calculated Voltages row. They should sum to 10 V.
6. Connect the circuit, and adjust the source to a value of 10 V.
7. Measure the voltage across each resistor, and record these in the Measured Voltages row in Table 10-1.
8. Add the measured voltages from step 7 and enter their sum in the appropriate area of the table.
9. Repeat steps 1 through 8 with a different resistor combination, and record all corresponding results in Table 10-2.

DATA FOR EXPERIMENT 10

TABLE 10-1

	R_1	R_2	R_3	R_4	R_T	Calculated Current I (mA)	
Measured Resistance Values							
Calculated Voltages	V_1	V_2	V_3	V_4	V_{sum}		
Measured Voltages	V_1	V_2	V_3	V_4	V_{sum}		

TABLE 10-2

	R_1	R_2	R_3	R_4	R_T	Calculated Current I (mA)	
Measured Resistance Values							
Calculated Voltages	V_1	V_2	V_3	V_4	V_{sum}		
Measured Voltages	V_1	V_2	V_3	V_4	V_{sum}		

NOTES

QUESTIONS FOR EXPERIMENT 10

1. Suppose in this experiment, R_1 = 1.8 kΩ, R_2 = 3.3 kΩ, R_3 = 4.7 kΩ, R_4 = 8.2 kΩ, and V_s = 10 V. Then the voltage V_1 should be close to

(   )   (a) 1 V   (b) 1. 8 V   (c) 2.6 V   (d) 90 mV

2. Suppose the conditions are as in question 1, and the following measurements are taken: V_1 = 0.97 V, V_2 = 1.77 V, and V_3 = 2.65 V. Then, V_4 should be close to

(   )   (a) 4.55 V   (b) 4.61 V   (c) 0 V   (d) 5.39 V

3. Continuing with question 2, what would be the value of the current in this instance?

(   )   (a) 0.55 mA   (b) 0.30 mA   (c) 1.22 mA   (d) 1.00 mA

4. If R_1, R_2, and R_3 were exactly nominal in value and R_4 was larger than nominal, then
   (a) V_4 will be smaller than nominal
   (b) V_1, V_2, and V_3 will be larger than nominal
   (c) all voltage drops will be smaller than nominal
   (d) V_4 will be larger than nominal, but V_1, V_2, and V_3 will be
(   )   smaller than nominal

5. Explain in your own words the effect on the total current and each voltage if a single resistor is smaller than its nominal (color-coded) value.

6. Is it possible to have some voltages less than nominal and some voltages greater than nominal simultaneously in such a circuit? If so, how?

11

VOLTAGE DIVIDERS

REFERENCE READING

Principles of Electric Circuits: Sections 5–7, 5–9.

RELATED PROBLEMS FROM *PRINCIPLES OF ELECTRIC CIRCUITS*

Chapter 5, Problems 37 through 45.

OBJECTIVE

To verify the Voltage Divider Theorem.

EQUIPMENT

dc power supply 0–10 V
DMM or VOM
Potentiometer: 10 kΩ
Resistors (±5%): 1.6 kΩ   2.7 kΩ
                     2.4 kΩ   3.3 kΩ

BACKGROUND

Voltage dividers find many applications in electronic circuits. The requirement for voltages of different values often arises and is normally accomplished by a series-resistive circuit, as shown in Figure 11-1. Voltages V_A, V_B, V_C, and V_D are potential differences measured with respect to *circuit ground*. (See *Principles of Electric Circuits,* Section 5–9, for a discussion of circuit ground.) Circuit ground in this case is an arbitrarily defined zero-volt reference point. Though the earth ground symbol is being used in Figure 11-1, this does not in fact mean that this point has to be connected to the earth. Ground is simply the negative side of the battery in this case. Because the current in each resistor is the same, the individual voltage across each resistor is proportional, by Ohm's Law, to the respective resistor value.

Figure 11-2 shows how a potentiometer can be used as a continuous voltage divider. For this experiment its value has been chosen to be 10 kΩ, which is exactly

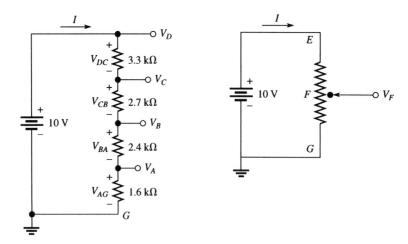

FIGURE 11-1   **FIGURE 11-2**

the sum of the four resistors in Figure 11-1. By appropriate positioning of the wiper arm, any of the voltages you obtain in the "discrete resistor" circuit of Figure 11-1 (and all values in between) can be derived. Potentiometers such as these are often used in this manner as volume controls in audio equipment.

PROCEDURE

1. For the circuit in Figure 11-1, calculate the nominal values for the voltage drops across each resistor—that is, V_{DC}, V_{CB}, V_{BA}, and V_{AG}—and record them in Table 11-1.
2. Use the information from step 1 to determine the nominal voltages V_A, V_B, V_C, and V_D and record them in Table 11-1. Note that all voltages are with respect to circuit ground.*
3. Calculate and record the circuit's total current I.
4. Construct the circuit, and then measure and record each of the voltages that you calculated in steps 1 and 2.
5. Complete the table by measuring the total current I and compare it with the expected value.
6. Dismantle the circuit in Figure 11-1 and construct the circuit in Figure 11-2. Place the DMM set to **dc VOLTS** between the wiper and the grounded (reference) end of the potentiometer.
7. Calculate the theoretically expected values for the minimum and maximum values of V_{FG} as the wiper is moved from one extremity to the other. V_{FG} is the voltage at F with respect to G, which is ground.
8. Measure and record in Table 11-2 the minimum and maximum voltages referred to in step 7.
9. Adjust the wiper of the potentiometer until the voltage V_{FG} is equal to 1.6 V (see first column of Table 11-3). Without disturbing it, remove the potentiometer and measure the resistance between points E and F, then F and G. Record these values in Table 11-3. Be sure to turn off the power first.
10. Repeat step 7 for the remaining voltages in Table 11-3.
11. Now compare the measured resistance values between the four wiper positions and ground in Figure 11-2 with those that appear between points A, B, C, and D and ground in Figure 11-1.

---

*The ground symbol is for reference purposes only. There is no need to physically connect the negative of the supply to an actual "ground."

Name _____   Date _____

DATA FOR EXPERIMENT 11

TABLE 11-1

Voltage	V_{DC}	V_{CB}	V_{BA}	V_{AG}	V_A	V_B	V_C	V_D	I
Nominal									
Measured									

TABLE 11-2

Voltage	$V_{FG\ min}$	$V_{FG\ max}$
Nominal		
Measured		

TABLE 11-3

Voltage	1.6 V	4 V	6.7 V	10 V
R_{FG}				
R_{EF}				

NOTES

Name _____ Date _____

QUESTIONS FOR EXPERIMENT 11

1. If all the resistance values in Figure 11-1 were stepped up by a factor of 10, the new voltages would be
   (a) 10 times their old values
   (b) unchanged
( )   (c) one-tenth their old values
2. When resistor tolerance of 5 percent is taken into account, the smallest expected value for V_C in Figure 11-1 is
( )   (a) 6.81 V   (b) 6.48 V   (c) 6.37 V   (d) 6.57 V
3. If the 2.4 kΩ resistor in Figure 11-1 were mistakenly replaced with a 240 Ω resistor, the voltages that would be affected would be
   (a) V_A and V_B only   (b) V_B only
( )   (c) V_A, V_B, V_C, and V_D (d) V_A, V_B, and V_C only
4. The current I in Figure 11-2 will change as the arm of the potentiometer is moved.
( )   (a) True   (b) False
5. What advantage does the potentiometer method have over the fixed resistor method for voltage dividing?

6. In step 9 you measured V_{FG} = 1.6 V. If you had measured V_{EF}, what would have been its expected value? Explain.

56

12

TROUBLESHOOTING IN SERIES CIRCUITS

REFERENCE READING

Principles of Electric Circuits: Sections 5–9 and 5–10.

RELATED PROBLEMS FROM *PRINCIPLES OF ELECTRIC CIRCUITS*

Chapter 5, Problems 51 through 56.

OBJECTIVE

To examine the effects of short- and open-circuited resistors in series circuits.

EQUIPMENT

dc power supply 0–12 V
DMM or VOM
Resistors (±5%):   2 kΩ (three)
                    10 Ω (one)
                    1 MΩ (one)

BACKGROUND

It may come as a surprise that very little theory is required to perform simple troubleshooting analysis on circuits. Armed with Ohm's Law and Kirchhoff's Voltage Law, simple faults such as short-circuited and open-circuited resistors can be diagnosed given only voltage information on the circuit in question. The analysis rests on two key ideas: an open resistor will carry no current, but will have the full open-circuit voltage across it; and a shorted resistor will have no voltage, but will carry the total series current through it.

In this experiment, you will investigate the effects of single isolated faults on a three-resistor voltage divider (Figure 12-1). Only one fault (short- or open-circuit) is introduced at a time, and resulting voltages are checked. Notice that current is

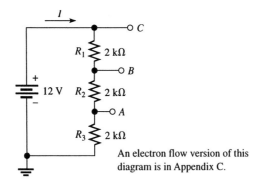

R_1 2 kΩ

12 V R_2 2 kΩ

R_3 2 kΩ

An electron flow version of this
diagram is in Appendix C.

FIGURE 12-1

not normally measured in troubleshooting exercises like this; the reason is that voltage can always be measured without breaking the circuit, whereas current measurement requires the breaking of the circuit.

A word about the simulation of shorts and opens in this experiment. To simulate a short circuit, we will replace one of the 2 kΩ resistors with a 10 Ω resistor. This resistance value is so much smaller than that of the 2 kΩ resistor that it can be assumed to be approximately a short circuit. To simulate an open, you can use a large-valued resistor such as 1 MΩ. This is fairly realistic, since it is often the case that resistors actually behave like a *very high resistance,* as opposed to a *perfect open,* when they go faulty in this way.

PROCEDURE

1. For the circuit in Figure 12-1, calculate the *normal* values for each of the voltages V_A, V_B, and V_C and record them in Table 12-1. Normal voltages mean those that you would expect if the circuit were operating properly. All voltages are with respect to circuit ground.
2. Construct the circuit and verify these voltages with the DMM. Record the measured values in Table 12-2. They should, of course, be close to the theoretical values; any deviation from these values can be attributed mostly to the tolerance of the resistors.
3. Consider now that R_1 (the uppermost resistor in Figure 12-1) becomes *short-circuit.* Calculate and record in the column headed R_1 S/C, the expected voltages at the points A, B, and C under this condition. *Recall that a short circuit has a resistance of 0 Ω.*
4. Now simulate a shorted R_1 by removing and replacing it with a 10 Ω resistor. Measure the fault voltages and record them in Table 12-2.
5. Repeat these calculations for each of the three resistors, entering all calculated data in the appropriate columns of Table 12-1.
6. Repeat this procedure with the remaining resistors. Don't forget to replace R_1 with the original 2 kΩ when you simulate a shorted R_2.
7. Consider now that R_1 becomes *open-circuit.* Calculate and record in the column headed R_1 O/C, the expected voltages at the points A, B, and C under this condition. *Recall an open circuit has a resistance of ∞ Ω and therefore no current flows in the circuit.*

8. You can simulate an open R_1 by removing and replacing it with the 1 MΩ resistor. Do this, recording the fault voltages in the appropriate area of Table 12-2.

9. Repeat these calculations for each of the three resistors, entering all calculated data in the appropriate columns of Table 12-1.

10. Compare the measured and calculated data. Apart from some minor deviations due to resistor tolerance and the fact that the simulations were not perfect opens and shorts, the data should compare well.

DATA FOR EXPERIMENT 12

TABLE 12-1 *Calculated data*

Voltages	Normal	Fault Voltages					
		R_1 S/C	R_2 S/C	R_3 S/C	R_1 O/C	R_2 O/C	R_3 O/C
V_A							
V_B							
V_C							

S/C means Short Circuit.
O/C means Open Circuit.

TABLE 12-2 *Measured data*

Voltages	Normal	Fault Voltages					
		R_1 S/C	R_2 S/C	R_3 S/C	R_1 O/C	R_2 O/C	R_3 O/C
V_A							
V_B							
V_C							

NOTES

QUESTIONS FOR EXPERIMENT 12

[NOTE: S/C = short circuit and O/C = open circuit.]

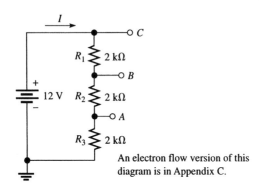

An electron flow version of this
diagram is in Appendix C.

FIGURE 12-2

1. Suppose there is a single fault in the circuit of Figure 12-2. You
    measure $V_C = 12$ V and $V_B = 12$ V. You conclude that
    (a) R_1 is S/C   **(b)** R_1 or R_2 or R_3 is S/C   **(c)** R_1 is O/C
    (d) R_1 is S/C or R_2 is O/C or R_3 is O/C

(   )

2. Suppose you measure the same voltages as in question 1, but in
    addition measure $V_A = 6$ V. You conclude that
    (a) R_2 is O/C   **(b)** R_3 is O/C   **(c)** R_1 is O/C   **(d)** R_1 is S/C

(   )

3. Suppose you measure the same voltages as in question 1, but in
    addition measure $V_A = 0$ V. You conclude that
    (a) R_1 or R_2 is O/C   **(b)** R_3 is S/C   **(c)** R_3 is O/C
    (d) R_2 is O/C

(   )

4. Suppose you measure the same voltages as in question 1, but in
    addition measure $V_A = 12$ V. You conclude that
    (a) R_3 is O/C   **(b)** R_2 is S/C   **(c)** R_1 is S/C   **(d)** R_3 is S/C

(   )

5. What possible single faults might cause V_C to be 12 V and V_B to
    be 6 V in the circuit of Figure 12-2?

6. When a resistor in a series circuit becomes short-circuit, this
    reduces the total resistance, and causes the current through the
    remaining resistors to increase. Point out the potential hazard
    of this increased current.

13

KIRCHHOFF'S CURRENT LAW

REFERENCE READING

Principles of Electric Circuits: Sections 6–1 through 6–3.

RELATED PROBLEMS FROM *PRINCIPLES OF ELECTRIC CIRCUITS*

Chapter 6, Problems 9 through 13.

OBJECTIVE

To verify Kirchhoff's Current Law for dc circuits.

EQUIPMENT

dc power supply 0–10 V
DMM or VOM
Resistors ($\pm 5\%$): 1.2 kΩ 4.7 kΩ
                                 1.8 kΩ 5.6 kΩ
                                 2.4 kΩ 6.8 kΩ
                                 3.3 kΩ 8.2 kΩ

BACKGROUND

Kirchhoff's Current Law states that the sum of the currents into a junction is equal to the sum of the currents out of that junction. With reference to Figure 13-1, this implies that

$$I_T = I_1 + I_2 + I_3 + I_4$$

If you connect up this circuit, you should be able to verify this rule by measuring each of the currents. Beginning students often have considerable difficulty in positioning the current meter in the right place for a particular current measurement. To help overcome this, when you connect up the circuit in Figure 13-1, use a piece

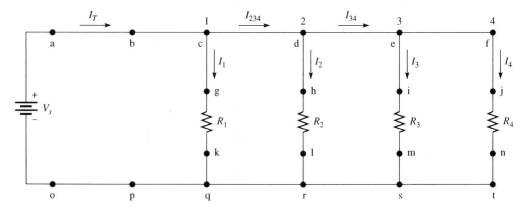

FIGURE 13-1

of hookup wire for each connection shown between heavy dots labeled 'a' through 'e' on the circuit diagram. This will make it easier in the procedure when deciding where to place the ammeter.

PROCEDURE

1. Select four resistors from the eight available and, with the power supply switched off, connect them in parallel, as shown in Figure 13-1. Beginning students often have a difficult time deciding where to place the ammeter to measure the various currents. To make this easier, you can use the connection method outlined in the next step. Otherwise go directly to step 3.

2. *Use separate pieces of hookup wire to make each of the connections between the heavy dots labeled 'a' through 't' in the circuit diagram.* You will require *eighteen* pieces of wire for this purpose. When you have completed this tedious process, you should have twenty distinct connection points, including those at the terminals of the power supply.

3. Record your selected resistor values in Table 13-1.

4. Set up the power supply for a voltage of 10 V at the output terminals.

5. Now, to measure current I_1, remove the piece of hookup wire that connects *junction 1* with the *upper end* of resistor R_1, and replace it with the DMM set to **dc mAMPS** on an appropriate range; see Figure 13-2. The ammeter is now

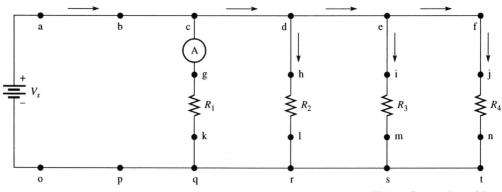

Electron flow versions of these
diagrams are in Appendix C.

FIGURE 13-2

in series with this resistor and should indicate the correct value of current. If you did not use the hookup wire in step 2, simply position the ammeter so that it measures the current in R_1. Measure and record this current.

6. Repeat this procedure for each of the remaining branch currents I_2, I_3, and I_4, recording all data in Table 13-1. In each case, don't forget to replace the meter with the small piece of hookup wire when you are finished.

7. Now repeat this procedure by measuring the currents at the lower end of each resistor; that is, insert the meter into each branch, and check that these measured currents agree with those you got when you went into the top of the branch. Record these data in the appropriate areas of Table 13-1.

8. The currents I_T, I_{234}, and I_{34} can be measured in a similar manner. That is, you can remove the hookup wire and replace it with the meter in each case. Record all data in Table 13-1 and, as a check, repeat these measurements at the lower end of the schematic diagram by removing the corresponding wires.

9. Complete the remainder of the table by finding each of the sums $I_3 + I_4$, $I_2 + I_3 + I_4$, I_T, etc. The calculated values should compare well with the corresponding measured currents. Any errors can be attributed to errors in the measuring instrument.

10. If time permits, repeat steps 1 through 7 for the remaining four resistors and record all data in Table 13-2.

DATA FOR EXPERIMENT 13

TABLE 13-1

	R_1	R_2	R_3	R_4	R_T		
Selected Nominal Values (kΩ)							
Measured Currents (mA) (top)	I_1	I_2	I_3	I_4	I_{34}	I_{234}	I_T
Measured Currents (mA) (bottom)	I_1	I_2	I_3	I_4	I_T	I_{234}	I_T
Calculated Currents by KCL	I_{34}		I_{234}		I_T		

TABLE 13-2

	R_1	R_2	R_3	R_4	R_T		
Selected Nominal Values (kΩ)							
Measured Currents (mA) (top)	I_1	I_2	I_3	I_4	I_{34}	I_{234}	I_T
Measured Currents (mA) (bottom)	I_1	I_2	I_3	I_4	I_T		
Calculated Currents by KCL	I_{34}		I_{234}		I_T		

NOTES

QUESTIONS FOR EXPERIMENT 13

1. Suppose that the following measurements are taken in the circuit in Figure 13-1: $I_T = 6.36$ mA, $I_1 = 2.13$ mA. Then the measured value of I_{234} should be in the vicinity of

(   )   (a) 8.49 mA   (b) 4.23 mA   (c) 4.48 mA   (d) 4.81 mA

2. Continuing with question 1, if in addition, I_2 were measured and found to be 1.79 mA, you would expect the measured value of I_{34} to be in the vicinity of

(   )   (a) 2.44 mA   (b) 4.23 mA   (c) 6.02 mA   (d) 4.48 mA

3. Continuing with question 2, if in addition, I_3 were measured and found to be 0.97 mA, you would expect the measured value of I_4 to be in the vicinity of

(   )   (a) 2.44 mA   (b) 4.23 mA   (c) 1.47 mA   (d) 0.97 mA

4. For measurement of *each* of the resistor currents in the circuit, how many possible positions for the ammeter are there?

(   )   (a) one only   (b) two   (c) three

5. The power supply current is the largest current in this circuit. Why?

6. In a parallel circuit like this the voltage across each branch is the same as the power supply. What does this imply about the resistances of the branch having the smallest current and the branch having the largest current?

PARALLEL RESISTIVE CIRCUITS

REFERENCE READING

Principles of Electric Circuits: Sections 6–1 through 6–5.

RELATED PROBLEMS FROM *PRINCIPLES OF ELECTRIC CIRCUITS*

Chapter 6, Problems 14 through 21.

OBJECTIVE

To verify that the total resistance in a parallel-connected circuit is given by the reciprocal rule.

EQUIPMENT

dc power supply 0–10 V
DMM or VOM
Resistors ($\pm 5\%$): 1.2 kΩ   2.4 kΩ
                   1.5 kΩ   3 kΩ
                     2 kΩ   1 MΩ

BACKGROUND

In a parallel circuit, each component has the same voltage across it, and the currents through each resistor are independent of one another. As more resistors are added in a parallel circuit, each one takes its own current from the supply independently of the others. As a result, the total current drawn increases. The power supply sees this as a reduction in the total resistance R_T of the circuit. In this experiment we are going to determine the effect of adding more resistors to a parallel connection, using the ohmmeter. We will then verify the total resistance formula for parallel (more than two) resistors, which is

$$1/R_T = 1/R_1 + 1/R_2 + 1/R_3 + \ldots + 1/R_n$$

We will then verify the simple product over sum rule for the special case of two resistors in parallel:

$$R_T = \frac{R_1 \cdot R_2}{(R_1 + R_2)}$$

You should realize that R_T will always be less than the smaller of R_1 and R_2. Then we will learn how to calculate the value of an unknown resistor in a two-resistor parallel connection when the values for one resistor and R_T are already known. The appropriate formula for this is

$$R_X = \frac{R_1 \cdot R_T}{(R_1 - R_T)}$$

where R_T is the total resistance, R_X is the unknown resistance, and R_1 is the known resistance of the pair.

Finally we will observe the adverse effects of the body resistance when measuring large-valued resistors with our fingers touching the ends of the resistor.

PROCEDURE

1. Measure the actual values of each of the five resistors (excluding the 1 MΩ) with the DMM on **OHMS,** and record in Table 14-1.
2. Using the *measured* values of these resistors, use the formula (the reciprocal rule) for R_T (see Background section) to determine the expected total resistance for each of the connections in Table 14-2. Note that an additional resistor is added each time you move to the right in the table. You will calculate four different combinations in all. (*Do not* use the color-coded values for these calculations.)
3. Insert the 3 kΩ resistor into your project board, and connect across it, with the DMM set on its **OHMS** function. The reading should correspond to the value you obtained for this resistor in step 1.
4. Now connect, in parallel, the 2.4 kΩ resistor, making up the first combination in Table 14-2, and watch the reading on the DMM. It should have decreased, indicating that the *total resistance* has decreased. It should also indicate a value in the vicinity of that predicted in step 2. Record this as the measured value in the table.
5. Continue to add, one at a time, the remaining three resistors in the sequence described in the table, recording the reading for each combination. The resistance should continue to decrease until you have the value in the last column of the table.
6. Take any two of the resistors from your selection of five, and calculate, using the product over sum rule, the total parallel resistance for this combination. Record your data in Table 14-3.
7. Verify your work in step 6 by parallel-connecting the resistors and measuring the total resistance with the DMM. Record the measured value.
8. Now have your instructor or laboratory partner conceal one of the five resistors so that you are unable to see the color code. This can easily be done with adhesive tape or typewriter correction fluid. Place this resistor in parallel with one of the (known) other resistors of the group of five. Write the *measured value* of the known resistor as R_1 (from step 1) in Table 14-4.

9. Measure the total resistance of the combination, and record this as R_T in Table 14-4.
10. Use the formula given in the Background section of this experiment to determine the value of the unknown resistor (R_X) and record this calculated value in the table.
11. Peel off the adhesive tape, or scratch off the correction fluid, and read the color-coded value of the concealed resistor. When its tolerance is taken into account, its value should be close to the value calculated in step 10. In fact, the predicted value from step 10 should compare well with the measured value of this resistor in step 1.
12. Take the 1 MΩ resistor, and connect it across the DMM on its **OHMS** function. When you do this, *deliberately support the ends of the resistor with your fingers*. Unless your body resistance is very high, the indicated value will probably be substantially different from 1 MΩ and well out of its tolerance range. Record this measured value in Table 14-5 under the With Body heading.
13. Remove your fingers from the ends of the resistor, and the indicated value should be much closer to the nominal or expected value of the resistor. Record this true value of R in Table 14-5 under the Without Body heading.

Name _____   Date _____

DATA FOR EXPERIMENT 14

TABLE 14-1

	Resistance Values (kΩ)				
Nominal	3	2.4	2	1.5	1.2
Measured					

TABLE 14-2

	Total Resistance of Combination R_T			
Combination (kΩ)	3 ‖ 2.4	3 ‖ 2.4 ‖ 2	3 ‖ 2.4 ‖ 2 ‖ 1.5	3 ‖ 2.4 ‖ 2 ‖ 1.5 ‖ 1.2
Calculated Resistance (kΩ)				
Measured Resistance (kΩ)				

Note: ‖ means "in parallel with."

TABLE 14-3

R_1	R_2	$R_{T\text{ calc}}$	$R_{T\text{ meas}}$

TABLE 14-4

R_1	R_T	R_X

TABLE 14-5

	With Body	Without Body
Measured Resistance (1 MΩ)		

NOTES

Name _____ Date _____

QUESTIONS FOR EXPERIMENT 14

( )

( )

( )

( )

1. Calculate the total resistance of the following parallel combination of resistors: 2.4 kΩ and 1.2 kΩ.
   (a) 3.6 kΩ   (b) 0.8 kΩ   (c) 1.7 kΩ   (d) 1.4 kΩ

2. If both of the resistors in question 1 are 10 percent tolerance, what would be the smallest expected value for the parallel combination?
   (a) 3.24 kΩ   (b) 0.72 kΩ   (c) 1.53 kΩ   (d) 0.76 kΩ

3. Continuing with question 2, what would be the largest expected value for the parallel combination?
   (a) 0.88 kΩ   (b) 3.96 kΩ   (c) 0.96 kΩ   (d) 1.4 kΩ

4. Continuing with questions 2 and 3, would it be correct to say that the tolerance of the combination is also 10 percent in this case?
   (a) Yes, it would.   (b) No, it would not.

5. In your own words, explain why the measured resistance of a large-valued resistor is affected when you are touching the ends of the resistor. Be sure to include why the indicated value is *less* than the true value.

6. Can you calculate your body resistance from finger to finger using the information in Table 14-4 and a formula given in the Background section of this experiment? Explain.

CURRENT DIVIDERS

REFERENCE READING

Principles of Electric Circuits: Section 6–7.

RELATED PROBLEMS FROM *PRINCIPLES OF ELECTRIC CIRCUITS*

Chapter 6, Problems 32 through 36.

OBJECTIVE

To verify the Current Divider Theorem.

EQUIPMENT

dc power supply 0–10 V
DMM or VOM
Resistors ($\pm 5\%$): 1.6 kΩ 2.7 kΩ
                        2.4 kΩ 3.3 kΩ

BACKGROUND

In experiment 11 we learned how a string of series-connected resistors can be used to divide a source voltage into smaller portions. The voltage divider allowed us to calculate the distribution of voltages in such circuits *without* the requirement for calculating the current. In a similar manner, the Current Divider Theorem allows us to calculate the magnitudes of the currents in a parallel circuit, using only the total current and the resistor values and not the voltage. It is particularly useful when the circuit is driven by a (constant) current source as opposed to a constant voltage source.

The formula for the current I_X in a specific resistor R_X of a parallel combination R_T, when the total current I_T entering the junction is known, can be written

$$I_X = \frac{R_T I_T}{R_X}$$

where R_T is the total parallel resistance.

In this experiment we will verify the Current Divider Theorem by adjusting the source for a specified *total* current and measuring the individual branch (resistor) currents. Remember that resistor tolerance will affect your results; if only one resistor is out of tolerance, *all* of the resistor currents will be affected.

PROCEDURE

1. For the circuit in Figure 15-1, assuming a total current of 10 mA, use the Current Divider Theorem to calculate the nominal values for the currents I_1, I_2, I_3, and I_4, and record them in Table 15-1.
2. Construct the parallel circuit, and with the DMM set to indicate the total circuit current, adjust the source voltage until the current I_T is 10 mA.
3. Now use the DMM to measure each of the resistor currents I_1 through I_4, and record these in the Measured row in Table 15-1.
4. Complete the table by calculating the source voltage required to sustain the 10 mA current into the parallel circuit. Set the DMM on **VOLTS** and confirm this voltage. Remember that resistor tolerance and meter error will account for any minor variations between the theoretical and measured values.

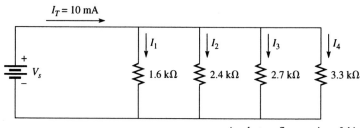

An electron flow version of this diagram is in Appendix C.

FIGURE 15-1

Name _____ Date _____

DATA FOR EXPERIMENT 15

TABLE 15-1

Currents (mA)	I_1	I_2	I_3	I_4	Voltage	V_s
Nominal					Nominal	
Measured					Measured	

NOTES

QUESTIONS FOR EXPERIMENT 15

1. If all the resistance values in Figure 15-1 were stepped up by a factor of 10 (and I remained constant at 10 mA), the new currents would be
   (a) 10 times their old values   (b) unchanged
   (c) one-tenth their old values

   ( )

2. Assuming I is adjusted to be 10 mA when resistor tolerance of 5 percent is taken into account, the smallest expected value for I_3 in Figure 15-1 is
   (a) 2.16 mA   (b) 1.99 mA   (c) 2.08 mA
   (d) none of these

   ( )

3. If the 2.4 kΩ resistor in Figure 15-1 were mistakenly replaced with a 24 kΩ resistor and I_T *were maintained constant*, the currents affected would be
   (a) $I_1, I_2, I_3,$ and I_4 (b) $I_2, I_3,$ and I_4 (c) $I_1, I_3,$ and I_4
   (d) I_2 only

   ( )

4. If all the resistors in Figure 15-1 were at the upper boundaries of their tolerance, the voltage V_S required to sustain 10 mA through the combination would be
   (a) 5.83 V   (b) 6.12 V   (c) 5.54 V   (d) none of these

   ( )

5. The smallest-valued resistance in a parallel connection of resistors will always take the largest current. Explain why this is so.

6. If the current I in Figure 15-1 is kept constant, comment on the effect of varying any single resistor on the branch currents $I_1, I_2, I_3,$ and I_4.

16

TROUBLESHOOTING IN PARALLEL CIRCUITS

REFERENCE READING

Principles of Electric Circuits: Section 6–10.

RELATED PROBLEMS FROM *PRINCIPLES OF ELECTRIC CIRCUITS*

Chapter 6, Problems 43 through 49.

OBJECTIVE

To examine the effect of short- and open-circuited resistors in parallel circuits.

EQUIPMENT

dc power supply 0–10 V
DMM or VOM
Resistors (±5%): 120 Ω* (2 W)
             20 kΩ
             30 kΩ
             62 kΩ
              1 MΩ

BACKGROUND

The purpose of this experiment is to investigate the effects of simple faults in parallel circuits. These simple faults will be "open" or "shorted" resistors, as they were in experiment 12, and again, only one fault will be introduced at a time. A word of caution when performing short-circuit tests in parallel circuits: The power supply must be short-circuit protected. Normally, this is accomplished by an adjustable

---

*A 120 Ω resistor is not required if the power supply has a current-limiting feature and/or short-circuit protection.

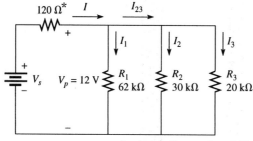

An electron flow version of this
diagram is in Appendix C.

FIGURE 16-1

current-limit feature; this control, when provided, will allow the user to adjust the maximum current capability of the power supply under any load conditions. In the event of a short circuit, the power supply will go into a *current limit* and the voltage across its terminals will drop to almost 0 V. In this experiment, you should adjust the current limit to a value of 100 mA. You can refer to the operator's manual or ask your instructor for the procedure required to do this.

  If the supply that you are using does not have a current-limit feature, then place a 120 Ω resistor in series with the power supply output, as shown in Figure 16-1. This resistor will limit the current to a little more than 100 mA when any of the parallel resistors goes short-circuit. When using this resistor, it is best to consider it as part of the power supply to monitor the voltage V_p (and not V_s) with the meter when setting up the voltage across the parallel combination. In this experiment, short circuits will be simulated by pieces of hookup wire; open circuits will be simulated by the 1 MΩ resistor.

PROCEDURE

1. For the circuit in Figure 16-1, calculate the nominal values for the currents I, I_1, I_{23}, I_2, and I_3, and record them in Table 16-1. Note that voltage across the parallel branches is given as 12 V in the figure.
2. If the supply does not have a current-limit feature, proceed to step 3; otherwise proceed as follows: Set the current limit to 100 mA. This is normally accomplished by *turning down* the voltage at the supply to 0 V, short-circuiting the output terminals, and then *turning up* the current control until the short-circuit current is 100 mA. A panel meter (on the front facia of the supply) is usually provided for measuring this current. *Do not* insert the 120 Ω resistor shown in Figure 16-1. Go directly to step 4. If in doubt, always insert the 120 Ω resistor.
3. If the supply does not have a current-limit feature, insert a 120 Ω resistor in series with the power supply terminals, as shown in Figure 16-1. This will limit the maximum current out of the supply to about 100 mA. Be sure to use a resistor with adequate power rating (at least 2 W).
4. Construct the parallel circuit, and adjust the source voltage until the *voltage across the parallel branches* is exactly 12 V. If the 120 Ω resistor is present, the actual terminal voltage at the supply will be slightly greater than 12 V.
5. Measure and record each of the currents listed in Table 16-2.

*Insert this resistor only if the power supply does not have a current-limit feature.

6. Consider now that R_1 were to go short-circuit. Calculate and record in Table 16-1 the expected values for the currents if this were to occur. Enter all values under the heading Shorted Resistors in the table. Remember that currents do not go to infinity, but are limited to approximately 100 mA by either the 120 Ω resistor or the current limit (V_s is only a little bigger than 12 V).

7. Simulate a shorted R_1 by inserting a short piece of hookup wire *in place of the resistor*. Measure and record the currents under these conditions. The measured values should compare well with theory.

8. Repeat these maneuvers for each of the resistors, recording the *expected* values of the currents in Table 16-1 and the *measured* values, after simulation of the required condition, in Table 16-2.

9. Consider now an open R_1. Calculate the expected values of the currents if R_1 were to fail *open-circuit* and record them in Table 16-1.

10. Simulate an open resistor by replacing R_1 with the 1 MΩ resistor, which, for all practical purposes, is close to an open circuit compared with the other resistors in this circuit. Measure all currents, and record in Table 16-2.

11. Repeat steps 9 and 10 for each of the resistors, recording the data in the respective tables.

Name _____ Date _____

DATA FOR EXPERIMENT 16

TABLE 16-1 *Calculated values*

Currents	Nominal Values (mA)	Shorted Resistors			Open Resistors		
		R_1	R_2	R_3	R_1	R_2	R_3
I							
I_1							
I_{23}							
I_2							
I_3							

TABLE 16-2 *Measured values*

Currents	Nominal Values (mA)	Shorted Resistors			Open Resistors		
		R_1	R_2	R_3	R_1	R_2	R_3
I							
I_1							
I_{23}							
I_2							
I_3							

NOTES

Name _____ Date _____

QUESTIONS FOR EXPERIMENT 16

[NOTE: S/C = short circuit and O/C = open circuit.]

1. Suppose a fault occurs in the circuit in Figure 16-1. You measure $I = 0.6$ mA. You conclude that
   (a) R_1 is O/C
   (b) R_1 and R_2 are O/C
   (c) R_1 *and* R_2 are O/C *or* R_3 is O/C
   (   ) (d) none of these

2. Suppose you measure the same current I as in question 1, but in addition measure $I_{23} = 0.4$ mA. You conclude that
   (a) R_1 and R_2 are O/C   (b) R_1 is O/C
   (   ) (c) R_2 is O/C   (d) R_3 is O/C

3. Suppose you measure the same current I as in question 1, but in addition measure $I_{23} = 0.6$ mA. You conclude that
   (a) R_1 and R_2 are O/C   (b) R_1 is O/C
   (   ) (c) R_2 is O/C   (d) R_3 is O/C

4. So long as any single resistor is shorted, I will have a value of approximately 100 mA regardless of the conditions of the remaining two resistors.
   (   ) (a) True   (b) False

5. What are the possible dangers of a resistor going short-circuit? Does it present any hazard to the remaining parallel resistors?

6. Why does a resistor, such as the 120 Ω series one, have to have such a large power rating when performing short-circuit tests as in this experiment?

17

SERIES-PARALLEL COMBINATIONS

REFERENCE READING

Principles of Electric Circuits: Sections 7–1, 7–2.

RELATED PROBLEMS FROM *PRINCIPLES OF ELECTRIC CIRCUITS*

Chapter 7, Problems 1 through 23.

OBJECTIVE

To examine total resistance in a series-parallel circuit.

EQUIPMENT

dc power supply 0–10 V
DMM or VOM
Resistors (± 5%):   2 kΩ
                    12 kΩ
                    24 kΩ

BACKGROUND

The most general kind of circuit connection is that which contains both series and parallel resistor configurations. The standard method of attack in this kind of problem is to first combine parallel resistors into a single equivalent, then add the series resistors.

The circuit in this experiment (Figure 17-1) contains a simple combination of two parallel resistors in series with a third resistor. Using Kirchhoff's Voltage Law, Kirchhoff's Current Law, and Ohm's Law, you can solve for all currents and voltages. In this case, you will calculate theoretical values for all of the quantities and verify them experimentally.

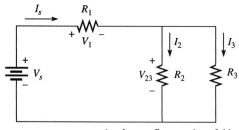

An electron flow version of this
diagram is in Appendix C.

FIGURE 17-1

PROCEDURE

1. Calculate the upper and lower limits of the resistance for each resistor and record these in Table 17-1.
2. Use the ohmmeter to measure the actual values of each resistor and record these in the R_{meas} row.
3. Complete Table 17-1 by calculating the minimum and maximum values for the total circuit resistance R_T of the circuit in Figure 17-1 and then measuring this resistance using an ohmmeter.
4. Using a value for V_s of 10 V, calculate the nominal, minimum, and maximum values for I_s in the circuit in Figure 17-1 and record them in Table 17-2.
5. Connect the circuit and adjust V_s to a value of 10 V. Measure the current I_s, and record it under I_{meas} in Table 17-2.
6. Measure the remaining currents and voltages and record them in Table 17-2.
7. Using a value for I_s of 0.5 mA, calculate the nominal, minimum, and maximum values for V_s in order to sustain this current through the series-parallel combination. Record these values in Table 17-3.
8. Connect the circuit and adjust V_s until $I = 0.5$ mA. Measure the voltage and record it in Table 17-3. (Does it appear consistent with $R_{T\text{ meas}}$?)
9. Measure the remaining currents and voltages and record them in Table 17-3.

Name _____ Date _____

DATA FOR EXPERIMENT 17

TABLE 17-1

R	$R_1 = 2\ \text{k}\Omega$	$R_2 = 12\ \text{k}\Omega$	$R_3 = 24\ \text{k}\Omega$	$R_T = 10\ \text{k}\Omega$
R_{min}				
R_{max}				
R_{meas}				

TABLE 17-2

	I_{nom}	
	I_{min}	
	I_{max}	
	I_{meas}	
$V_s = 10\ \text{V}$	$I_{2\ meas}$	
	$I_{3\ meas}$	
	$V_{1\ meas}$	
	$V_{23\ meas}$	

TABLE 17-3

	$V_{s\ nom}$	
	$V_{s\ min}$	
	$V_{s\ max}$	
$I = 0.5$ mA	$V_{s\ meas}$	
	$I_{2\ meas}$	
	$I_{3\ meas}$	
	$V_{1\ meas}$	
	$V_{23\ meas}$	

NOTES

QUESTIONS FOR EXPERIMENT 17

1. In Figure 17-1, if R_2 were less than nominal, then for a given voltage V_s,
   (a) I_2 would be less than nominal
   (b) I_3 would be less than nominal
   (c) I_s would be less than nominal

( ) (d) I_3 would be greater than nominal

2. If an additional resistor R_4 were added in parallel with R_1, then the total resistance R_T would

( ) (a) decrease   (b) increase   (c) not change

3. In this circuit, if the value of R_3 were doubled, then for a given voltage V_s,
   (a) I_3 would halve   (b) I_s and I_3 would halve
   (c) I_2 would be unaffected

( ) (d) all currents would be affected

4. If all of the resistors in this circuit have a tolerance of ± 5 percent, then the tolerance of R_T will be

( ) (a) $\pm 5\%$ (b) $\pm 7.5\%$ (c) $\pm 15\%$ (d) none of these

5. Decreasing any resistor value in such a circuit will always decrease the total resistance R_T. Is this true? Explain.

6. How many different possible values of R_T are obtainable with the three resistors if each circuit is to be a series-parallel combination of the kind in Figure 17-1? Calculate all of the possible values using the resistors in this experiment.

VOLTAGE DIVIDERS
WITH RESISTIVE LOADS

REFERENCE READING

Principles of Electric Circuits: Section 7–3.

RELATED PROBLEMS FROM *PRINCIPLES OF ELECTRIC CIRCUITS*

Chapter 7, Problems 25 through 34.

OBJECTIVE

To examine the effects of loads attached to voltage dividers.

EQUIPMENT

dc power supply 0–10 V
DMM or VOM
Resistors ($\pm 5\%$): 100 kΩ
                4.7 kΩ (two)
                10 kΩ (two)
                1 kΩ (three)

BACKGROUND

In experiment 11 you examined the basic voltage divider. It consisted of a string of series resistors with values chosen to yield particular voltage levels. Because you had not yet studied series-parallel combinations, we were unable to study the effects of loads connected to the points when the voltages were available. When loads are connected in this way (Figure 18-1), the situation is referred to as a *loaded voltage divider*.

When a load is connected, however, the voltage at the point in question changes. To offset this change, dividers have to be designed with a knowledge of the loads so that the correct voltage is achieved *after* the load is connected. Alterna-

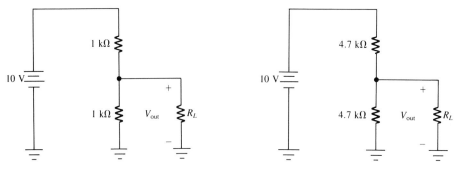

FIGURE 18-1   **FIGURE 18-2**

tively, the divider chain resistor values can be made very small, compared with those of the load resistors. (This is referred to as a *lightly loaded divider.*)

In this experiment, you will examine the effect of both light and heavy loads connected to divider chains. You will have to do a fair amount of series-parallel resistor analysis, so make sure you are proficient before going ahead.

PROCEDURE

1. For the voltage divider shown in Figure 18-1, calculate the expected value of V_{out}, assuming that there is no load (i.e., $R_L = \infty\ \Omega$ = an open circuit) connected to the divider. Record this in Table 18-1 in the Calculated column.
2. Calculate the new value of V_{out} when it is loaded with a resistor R_L equal to 100 $k\Omega$. You must first combine the 1 $k\Omega$ and 100 $k\Omega$ resistors in parallel, then add them to the series 1 $k\Omega$ to find the total resistance. The (loaded) value of V_{out} can then easily be determined by using the Voltage Divider Theorem.
3. Repeat the procedure in step 2 with each of the remaining load resistors, 10 $k\Omega$ and 1 $k\Omega$. You should find that in each case, the voltage V_{out} is less than in the previous case.
4. Connect the circuit, and measure the *unloaded* value of V_{out}. Then attach the load resistors listed in Table 18-1, one at a time, and measure V_{out} for each case. Record all data in the Measured column of Table 18-1.
5. Repeat the above steps for the second divider, shown in Figure 18-2, and record the results in Table 18-2. Notice that the loads are the same; only the resistors in the voltage divider have been changed. In this case, the loaded voltages are not as close to the unloaded voltage as they were in the first figure. You should try to appreciate why this is the case before you turn to the questions for this experiment.

Name _____ Date _____

DATA FOR EXPERIMENT 18

TABLE 18-1

R_L (kΩ)	V_{out}	
	Calculated	Measured
Unloaded		
100		
10		
1		

TABLE 18-2

R_L (kΩ)	V_{out}	
	Calculated	Measured
Unloaded		
100		
10		
1		

NOTES

QUESTIONS FOR EXPERIMENT 18

1. When a load is connected to a voltage divider in the manner shown in Figure 18-1,
    (a) the value of V_{out} always decreases from its no-load value
    (b) the value of V_{out} always increases from its no-load value
( )   (c) the value of V_{out} does not change
2. A light load (R_L large) affects V_{out} more than a heavy load (R_L small).
( )   (a) True   (b) False
3. When the load is connected across the lower resistor in Figure 18-1, the voltage across the upper resistor
( )   (a) is unaffected   (b) increases   (c) decreases
4. If a 10 kΩ load were connected to the output of Figure 18-2, the voltage V_{out} would fall to
( )   (a) 6.80 V   (b) 0.81 V   (c) 9.19 V   (d) 4.05 V
5. When loads are connected to voltage dividers, the total dc current drawn from the source changes. Explain why this occurs and why it always increases.

6. The loaded voltages differed from the unloaded voltage substantially more in Figure 18-2, with the larger divider resistors, than they did in Figure 18-1, with the smaller divider resistors. Explain why this was the case.

VOLTMETER LOADING

REFERENCE READING

Principles of Electric Circuits: Section 7–4.

RELATED PROBLEMS FROM *PRINCIPLES OF ELECTRIC CIRCUITS*

Chapter 7, Problems 35 through 38.

OBJECTIVE

To determine the circuit-loading effect of a voltmeter.

EQUIPMENT

dc power supply 0–12 V	Resistors ($\pm 5\%$, two each): 2 kΩ
VOM (dc sensitivity 20 kΩ/V)	20 kΩ
DMM ($>$10 MΩ input impedance)	200 kΩ

BACKGROUND

It is not possible to make a measurement in an electrical circuit without affecting the outcome of that measurement. In fact, this principle applies to measurements of any kind in science. In electronic circuits this effect is known as *loading*. In the case of measuring current, the ammeter resistance adds to the original resistance in the circuit, causing us to underread the actual current. When measuring voltage, the addition of the voltmeter appears as a parallel resistance to the element in question, causing us to underread the voltage.

The extent to which the meter affects the measured voltage depends on the magnitude of the resistance of the instrument. For this reason, the instrument should have the highest resistance possible—it should not affect the circuit in question when placed in parallel. All meters, however, have some resistance. The VOM is perhaps the worst, having a typical resistance of 20 kΩ per volt of range on the

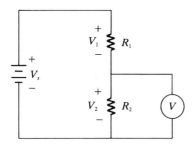

FIGURE 19-1

scale. On the other hand, a good DMM might have a resistance as high as 10 MΩ regardless of range. Clearly, the best measurements are given by the DMM, whereas those from the VOM can be inaccurate depending on the resistances in the circuit under test.

In this experiment you will learn how to determine the loading effect of a voltmeter. Using the simple two-resistor circuit in Figure 19-1, you will use both the VOM and DMM to measure the voltage V_2. As the value of R_1 and R_2 are increased, you should see the loading effect of the VOM become more pronounced; the DMM will yield more accurate results, even when $R_1 = R_2 = 200$ kΩ.

PROCEDURE

1. For the circuit in Figure 19-1, for series 2 kΩ resistors, ignoring resistor tolerance, calculate the voltages V_1 and V_2 that would be indicated by *ideal* voltmeters positioned to measure these voltages. Record the results of these calculations in Table 19-1 in the Ideal row.
2. Inspect the VOM, and decide which is the most suitable range for measurement of voltages in the vicinity of those you calculated in step 1. Use the VOM sensitivity figure in Ω/V (usually printed at the base of the meter scale) to calculate the resistance of the VOM on this range. Most VOMs have a sensitivity of 20 kΩ/V on their dc ranges, but see your instructor if you are unsure of this figure.
3. Using the calculated meter resistance from step 2, calculate the voltages V_1 and V_2 that you would expect this VOM to indicate when properly positioned to measure these voltages. Remember when the VOM resistance is taken into account, the circuit becomes one of the series-parallel type (the VOM resistance is in parallel with the resistance whose voltage is being measured). Record these values in the VOM (expected) row in Table 19-1.
4. Construct the circuit with the 2 kΩ resistors, and use the VOM to measure both V_1 and V_2. Record this measured data in the VOM (measured) row in Table 19-1.
5. Now repeat step 4 with the DMM. The DMM has a much higher resistance than the VOM and should indicate voltages closer to the truth. Record the voltages indicated by the DMM in Table 19-1 in the DMM row.
6. Repeat the above steps for resistors of 20 kΩ and then 200 kΩ, recording all data in Tables 19-2 and 19-3. The VOM resistance is the same for all of these measurements so long as you do not move from the original range chosen in step 2.

Name _____ Date _____

DATA FOR EXPERIMENT 19

TABLE 19-1

$R_1 = R_2 = 2\ \text{k}\Omega$	V_1	V_2
Ideal		
VOM (expected)		
VOM (measured)		
DMM		

TABLE 19-2

$R_1 = R_2 = 20\ \text{k}\Omega$	V_1	V_2
Ideal		
VOM (expected)		
VOM (measured)		
DMM		

TABLE 19-3

$R_1 = R_2 = 200\ \text{k}\Omega$	V_1	V_2
Ideal		
VOM (expected)		
VOM (measured)		
DMM		

NOTES

Name _____ Date _____

QUESTIONS FOR EXPERIMENT 19

1. The greatest error due to voltmeter loading occurs in
   (a) high-resistance circuits
   (b) low-resistance circuits
( )   (c) neither of these

2. In the circuit in Figure 19-1, the measurement error depends on
   (a) R_2 and meter resistance only   (b) R_2 only
( )   (c) R_1 and R_2 only   (d) R_1, R_2, and meter resistance

3. When the VOM is placed across R_2, the voltage across R_1
( )   (a) increases   (b) decreases   (c) does not change

4. When the meter resistance is equal to the resistance measured across, the error in Figure 19-1 is
   (a) 100%   (b) 50%   (c) 33⅓%
( )   (d) insufficient information

5. When a VOM is placed across either of the resistors in Figure 19-1, it affects the total resistance of the circuit, the total current drawn by the circuit, and both voltages. In which direction do each of these quantities change?

6. The typical VOM sensitivity is 20 kΩ/V on the dc range. However, a typical DMM has a constant resistance of 10 MΩ on any range. Assuming these figures, on what voltage range does the resistance of a typical VOM equal that of a typical DMM?

LADDER NETWORKS

REFERENCE READING

Principles of Electric Circuits: Section 7–5.

RELATED PROBLEMS FROM *PRINCIPLES OF ELECTRIC CIRCUITS*

Chapter 7, Problems 39 through 45.

OBJECTIVE

To become familiar with ladder networks.

EQUIPMENT

dc power supply 0–12 V
DMM or VOM
Resistors (±5%)   2 kΩ (two)
              2.4 kΩ
               3 kΩ
             3.6 kΩ
             12 kΩ
            10 kΩ (five)
           20 kΩ (three)

BACKGROUND

Ladder networks are a special type of series–parallel network that commonly occur in a variety of electronic circuitry. These networks are easily analyzed using a

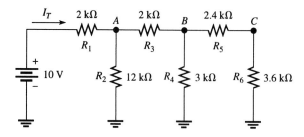

FIGURE 20-1

"backwards-approach" as outlined in Floyd's section 7–5. For the network shown in Figure 20-1, the voltages and currents can easily be determined once the total resistance R_T seen by the source is known. Starting at the far right end of the circuit, find R_{BG}, the total resistance from point B to ground. This is the series combination of R_5 and R_6 in parallel with R_4. Then combine this in series with R_3 and then this equivalent resistance in parallel with R_2 to find R_{AG}, the resistance from point A to ground. Finally, add the series combination of R_{AG} to R_1 and find R_T. It is now a simple matter to compute the total current I_T and then the voltages of A, B, and C with respect to circuit ground. The $R/2R$ ladder network in Figure 20-2 is analyzed in the same manner. These $R/2R$ networks are used in the process of digital-to-analog voltage conversion.

PROCEDURE

Part A: The General Ladder Network

1. For the circuit given in Figure 20-1, assuming nominal values for the resistors, calculate the nominal voltages V_A, V_B, and V_C all measured with respect to circuit ground. Record the data in Table 20-1.
2. Construct the circuit and measure the voltages with the DMM. Record the measured values in Table 20-1. The voltages should be close to your theoretically computed voltages in step 1.

Part B: The *R/2R* Ladder Network

1. For the $R/2R$ ladder network given in Figure 20-2, assuming nominal values for the resistors, calculate the nominal voltages V_A, V_B, and V_C all measured with respect to circuit ground. Record the data in Table 20-2. Record also the total resistance as seen by the source.
2. Construct the circuit and measure the voltages with the DMM. Record the measured values in Table 20-2. The voltages should be close to your theoretically computed voltages in step 1.

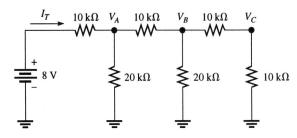

FIGURE 20-2

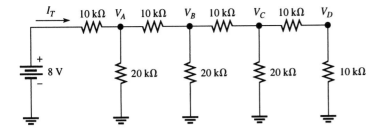

FIGURE 20-3

3. Replace the last resistor in the chain (the last 10 kΩ resistor) with a 20 kΩ resistor and add an additional pair of 10-kΩ resistors to produce the circuit in Figure 20-3.
4. Again, using the nominal values for the resistors, calculate the nominal voltages V_A, V_B, V_C and the additional voltage V_D all with respect to ground. Record all data in Table 20-3. Record also the total resistance as seen by the source.
5. Construct the circuit and measure the voltages with the DMM. Record the measured values in Table 20-3. The voltages should be close to your theoretically computed voltages in step 1.

DATA FOR EXPERIMENT 20

TABLE 20-1

Voltages	V_A	V_B	V_C
Nominal			
Measured			

TABLE 20-2

Voltages	V_A	V_B	V_C
Nominal			
Measured			
R_T			

TABLE 20-3

Voltages	V_A	V_B	V_C	V_D
Nominal				
Measured				
R_T				

NOTES

QUESTIONS FOR EXPERIMENT 20

1. In Figure 20-1, what is the approximate total circuit resistance as seen by the source?

( )   (a) 2 kΩ   (b) 3 kΩ   (c) 18 kΩ   (d) 5 kΩ

2. What is the approximate value of the current through the 3.6 kΩ resistor in Figure 20-1?

( )   (a) 1 mA   (b) 2 mA   (c) 0.5 mA   (d) 3 mA

3. The values of the nominal currents in each of the 20 kΩ resistors in Figure 20-2 are the same.

( )   (a) True   (b) False

4. If each of the resistors in the $R/2R$ ladder network in Figure 20-2 were doubled in value, while the source voltage were held at 8 V, then

(a) the voltages V_A, V_B, and V_C would stay the same

(b) the voltages V_A, V_B, and V_C would double

( )   (c) the voltages V_A, V_B, and V_C would halve

5. Explain in your own words why it is that the total circuit resistance in Figures 20-2 and 20-3 remains unchanged even though the circuit in Figure 20-2 has had resistors added to it.

6. Show how you would alter the $R/2R$ ladder network in Figure 20-3 to be able to obtain five output voltages of 4 V, 2 V, 1 V, 1/2 V and 1/4 V. Explain fully what needs to be done and draw a complete circuit diagram.

BALANCED AND UNBALANCED BRIDGE CIRCUITS

REFERENCE READING

Principles of Electric Circuits: Section 7–6.

RELATED PROBLEMS FROM *PRINCIPLES OF ELECTRIC CIRCUITS*

Chapter 7, Problems 46 through 48.

OBJECTIVE

To become familiar with the Wheatstone bridge circuit.

EQUIPMENT

dc power supply 0–12 V	Resistors (±5%): 1 kΩ (two)
DMM or VOM	10 kΩ (two)
Potentiometer 1 kΩ (ten-turn	2.2 kΩ
preferred) or decade box	3.3 kΩ
	4.7 kΩ

BACKGROUND

The bridge circuit, or Wheatstone bridge, as it is better known, finds wide application in instrumentation and measurement systems. A simple circuit, containing basically four or possibly five resistors, it can be used to demonstrate a variety of circuit principles. In this experiment, we will analyze and build a simple bridge circuit and demonstrate how it is balanced. Referring to Figure 21-1, *with respect to circuit ground,* the voltages at points C and D are given by

$$V_C = V_s \frac{R_3}{R_1 + R_3}$$

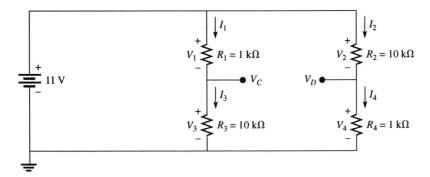

FIGURE 21-1

and

$$V_D = V_s \frac{R_4}{R_2 + R_4}$$

Here we used the Voltage Divider Theorem because each pair of resistors, R_1-R_3 and R_2-R_4, are series-connected.

The voltage V_{CD} across the *bridge points* is equal to the difference between these two voltages: $V_C - V_D$. The balance condition occurs when $V_C = V_D$ and therefore $V_C - V_D = 0$ V. Putting the equations for V_C and V_D equal to one another, we have for the balanced bridge:

$$R_1 R_4 = R_2 R_3$$

Under this condition, the potential difference between the bridge points is zero, and if a resistor or ammeter were connected between these points, no current would flow. The currents I_1 and I_3 are then equal, as are the currents I_2 and I_4.

PROCEDURE

Part A: Approximate Balanced Bridge

1. For the circuit given in Figure 21-1, assuming nominal values for the four resistors, calculate the currents I_1, I_2, I_3, and I_4; the voltages V_1, V_2, V_3, and V_4; and the voltages (with respect to circuit ground) V_C and V_D. Calculate also the voltage V_{CD}; that is, the voltage at point C with respect to point D. Enter all data in Table 21-1.
2. Before building the circuit, try to match, as *closely as possible,* the values of the two 10 kΩ and two 1 kΩ resistors. It does not matter if these resistors are not exactly equal to their nominal (coded) values, only that *corresponding resistors* are close in value.
3. Construct the circuit, and using the DMM on an appropriate function and range, measure and record all of the quantities calculated in step 1. The measured values should compare well with theory.
4. Exchange the positions of the two series resistors on the right side of the bridge; that is, bring the 10 kΩ down and move the 1 kΩ up to the top. This should give a reasonably good balance condition. (This will depend on how closely you matched the resistor values in step 2.) Recalculate all of the currents and voltages and record the data in Table 21-2.
5. Use the DMM to measure all of the currents and voltages except V_{CD} in the circuit, and record all data in Table 21-2.

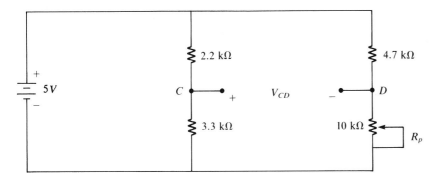

FIGURE 21-2 Note: A decade box may be used in place of the potentiometer.

6. Position the DMM to measure the voltage V_{CD} (across the bridge). Set it to progressively smaller ranges until you have the most precise reading (i.e., the most number of significant figures in the display). Record this measured voltage in Table 21-1. The voltage V_{CD} may not be exactly zero. Why not?

7. Because the voltage across the bridge is almost zero, any connection between the bridge points should result in no current flow. As a test of this condition, set the meter to a sensitive current (mA) range, and place it directly across points C and D. Because the resistance of an ammeter is very small, you are essentially placing a near short circuit across the bridge. The current that flows should be very small.

Part B: Exact Balancing of the Bridge

1. Using the DMM on **OHMS,** measure the *actual values* of the resistors that are to go into the circuit in Figure 21-2. Record the measured values in Table 21-3.

2. On the basis of *actual values* of the resistors, calculate the exact value for the potentiometer setting to achieve a bridge balance condition (V_{CD} = 0 V). Record this in the column headed R_p for Balance in Table 21-3.

3. Connect the wiper of the potentiometer to one of its ends and, using the DMM, set it so that its resistance is maximum ($\approx$10 kΩ).

4. With the power supply off, construct the circuit, and insert the DMM set to a large dc volts range (for example, 10 V or 20 V). You should place the DMM so that its common lead is connected to point C. It will then read a positive voltage when you switch on the power supply.

5. Switch on the power supply and adjust for 5 V. Slowly adjust the potentiometer until the voltage across the bridge points reads approximately 0 V.

6. Switch the DMM to a smaller voltage range (say 1 V or 2 V), and further adjust the potentiometer until you have approximately 0 V indicated.

7. You can get an even closer balance condition by switching to an mA range and further adjusting the potentiometer until you read 0 mA.

8. Before you dismantle the circuit, with the DMM across the bridge points, increase the power supply voltage by a few volts, and watch the reading on the meter. Any small out-of-balance voltage will be magnified and you can further adjust the potentiometer until you read closer to 0 V.

9. Remove the potentiometer, and measure its value. It ought to be close to the value you calculated in step 2. Record the measured value of the potentiometer setting in Table 21-3 under the column R_p for Balance.

Name _____ Date _____

DATA FOR EXPERIMENT 21

TABLE 21-1

Quantities	I_1	I_2	I_3	I_4	V_1	V_2	V_3	V_4	V_C	V_D	V_{CD}
Calculated											
Measured											

TABLE 21-2

Quantities	I_1	I_2	I_3	I_4	V_1	V_2	V_3	V_4	V_C	V_D	V_{CD}
Calculated											
Measured											

TABLE 21-3

Resistors	R_1	R_2	R_3	R_p for Balance	
Nominal (kΩ)	2.2	4.7	3.3	Calculated	
Measured (kΩ)				Measured	

NOTES

QUESTIONS FOR EXPERIMENT 21

1. In Figure 21-1, taking resistor tolerance (± 5 percent) into account, what is the *largest* expected value you could get for V_C in Figure 21-1?

( ) **(a)** 10.050 V   **(b)** 10.087 V   **(c)** 9.959 V   **(d)** 11.000 V

2. Again taking resistor tolerance into account, what is the *smallest* expected value you could get for V_D in Figure 21-1?

( ) **(a)** 0.950 V   **(b)** 0.913 V   **(c)** 1.000 V   **(d)** 0.864 V

3. When the bridge of Figure 21-1 is almost balanced (after exchanging the 1 kΩ and 10 kΩ resistors), what is the worst case (maximum) voltage you can expect to measure across the bridge points (C and D) when resistor tolerance is taken into account?

( ) **(a)** 0.182 V   **(b)** 0.087 V   **(c)** 0.095 V   **(d)** 0.055 V

4. In Figure 21-2, when perfectly balanced, the voltage across any resistor connected between the bridge points is
   (a) zero
   (b) dependent on the value of the resistor connected

( ) **(c)** equal to 5 V

5. Explain in your own words why, when you reverse the 10 kΩ and 1 kΩ resistors in step 4, the bridge is not perfectly balanced.

6. Explain in your own words the effect on V_C, V_D, and V_{CD} of making the potentiometer resistance in Figure 21-2 larger in value.

22

TROUBLESHOOTING
IN SERIES–PARALLEL CIRCUITS

REFERENCE READING

Principles of Electric Circuits: Section 7–7.

RELATED PROBLEMS FROM *PRINCIPLES OF ELECTRIC CIRCUITS*

Chapter 7, Problems 49 through 54.

OBJECTIVE

To examine the effects of short- and open-circuited resistors in series-parallel circuits.

EQUIPMENT

dc power supply 0–12 V
DMM or VOM
Resistors ($\pm 5\%$): 2 kΩ (two)
                           3 kΩ (five)
                  10 Ω
                  1 MΩ

BACKGROUND

Once again, it's time to take the knowledge gained in the last few experiments and use it to troubleshoot simple series-parallel circuits containing shorts and opens.

For simplicity and to avoid arithmetic calculations that will detract from the main purpose of the experiment, the circuit is made up of resistor values that easily combine in parallel and series. Consequently, it should be a simple matter to determine the potentials V_A, V_B, and V_C by inspection.

The simulated open and shorted conditions can be applied to each of four resistors in this case, and depending on where these "faults" occur, the voltages V_A, V_B, and V_C will be affected differently.

The questions at the end of the experiment reverse the situation. You are given symptoms and must diagnose the possible fault(s) that would give rise to them.

PROCEDURE

1. For the circuit in Figure 22-1, calculate the nominal values for the voltages V_A, V_B, and V_C, and record them in Table 22-1. All voltages are with respect to ground. Notice that the circuit is drawn with ground symbology like those in the references.
2. Construct the circuit; then measure and record the voltages V_A, V_B, and V_C. Record as measured nominal data in Table 22-2.
3. Consider now a shorted R_1. Calculate and record the resulting voltages V_A, V_B, and V_C if this were to occur. Enter the calculated values in the first column of Table 22-1 under the heading Shorted Resistors. Repeat this for each resistor in turn. Note R_1 equals 4 kΩ (2 + 2 kΩ).
4. Consider now removing R_1. Calculate and record the resulting voltages V_A, V_B, and V_C if this were to occur. Enter the calculated values in the first column of Table 22-1 under the heading Open Resistors. Repeat this for each resistor in turn.
5. Verify your calculations in steps 3 and 4 by simulating an open resistor with the 1 MΩ and a shorted resistor with the 10 Ω. Remember that because these resistors are *not ideal* shorts or opens, nor are the remaining resistors exactly equal to their nominal values, the results will differ slightly from the expected (theoretical) values.

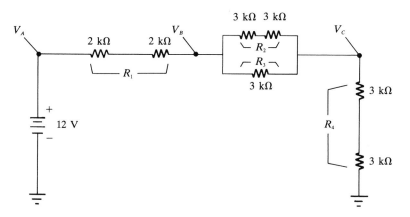

FIGURE 22-1

DATA FOR EXPERIMENT 22

TABLE 22-1 *Calculated data*

Voltages	Nominal	Shorted Resistors				Open Resistors			
		R_1	R_2	R_3	R_4	R_1	R_2	R_3	R_4
V_A									
V_B									
V_C									

TABLE 22-2 *Measured data*

Voltages	Nominal	Shorted Resistors				Open Resistors			
		R_1	R_2	R_3	R_4	R_1	R_2	R_3	R_4
V_A									
V_B									
V_C									

NOTES

QUESTIONS FOR EXPERIMENT 22

1. Suppose there is a single fault in the circuit of Figure 22-1. You measure $V_A = 12$ V and $V_B = 12$ V. You conclude that
   (a) R_1 is S/C   **(b)** R_3 or R_2 is O/C   **(c)** R_1 is O/C
   (d) R_1 is S/C or R_4 is O/C

( )

2. Suppose you measure the same voltages as in question 1, but in addition measure $V_C = 9$ V. You conclude that
   (a) R_1 is S/C   **(b)** R_3 is O/C   **(c)** R_2 is O/C
   (d) R_4 is O/C

( )

3. Suppose there is a single fault in the circuit, and you measure $V_A = 12$ V and $V_B = 7.2$ V. You conclude that
   (a) R_2 is S/C   **(b)** R_3 is S/C   **(c)** R_3 is O/C
   (d) R_2 or R_3 is S/C

( )

4. Suppose you measure $V_C = 0$ V. You conclude that
   (a) R_4 is S/C   **(b)** R_1 is O/C or R_4 is S/C
   (c) R_2 or R_3 is O/C   **(d)** R_4 is O/C

( )

5. What is the highest voltage to which V_C can rise, and under what single fault condition will this occur?

6. If one of the resistors becomes a short-circuit, which one will cause the greatest percentage increase in supply current?

23

LOADING OF THE VOLTAGE SOURCE

REFERENCE READING

Principles of Electric Circuits: Section 8–1.

RELATED PROBLEMS FROM *PRINCIPLES OF ELECTRIC CIRCUITS*

None.

OBJECTIVE

To examine the behavior of ideal and real voltage sources under loaded conditions.

EQUIPMENT

dc power supply 0–10 V
DMM or VOM
Resistors (±5%): 100 Ω   2.0 kΩ
200 Ω   2.7 kΩ
470 Ω   3.6 kΩ
620 Ω   4.7 kΩ
820 Ω   6.2 kΩ
1.1 kΩ   8.2 kΩ
1.5 kΩ

BACKGROUND

The voltage measured across the terminals of a battery or dc power supply is generally referred to as the *open-circuit voltage*. The amount by which it changes may be negligibly small, as in a regulated power supply, or fairly large, as in a dry cell with heavy load. The change itself is due to internal resistance, which tends to "absorb" some of the open-circuit voltage when current has to pass through it, as is the case when the source is loaded. The internal resistance is modeled as a resistor of value R_s to make it easier to see the voltage divider action that occurs.

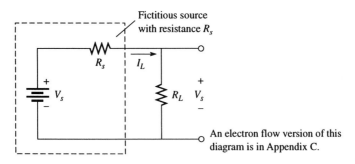

FIGURE 23-1

In this experiment you will examine the terminal voltage/current characteristics of supplies with differing values of R_s. The regulated power supply (which for most practical purposes has a value of $R_s = 0 \ \Omega$) and a resistor can serve to model a real, nonregulated source. The open-circuit voltage is the value measured under no-load conditions ($R_L = \infty \ \Omega$); then as the load becomes heavier ($R_L \downarrow$), the terminal voltage will change. Remember the terminal voltage for you is that measured across the load resistor, *not* across the source itself.

The terminal characteristic of an unregulated source such as this can be described by a straight-line equation of the form $y = mx + b$. In this context, y is replaced by V_t, the terminal voltage; x is replaced by I_L, the load current; and the slope of this line m is, in fact, the negative of the internal or source resistance $-R_s$. The equation is developed as follows.

Application of Kirchhoff's Voltage Law to the circuit in Figure 23-1 yields

$$V_t = V_s - (\text{voltage drop across } R_s)$$
$$= V_s - I_L R_s$$

which, written in normal straight-line form, is

$$V_t = \underset{\underset{\text{slope}}{\uparrow}}{-R_s I_L} + \underset{\underset{\text{intercept}}{\uparrow}}{V_s}$$

PROCEDURE

1. For the circuit in Figure 23-1, calculate the terminal voltage V_t and load current I_L for each value of R_L in Table 23-1. Do this for each value of source resistance in turn.
2. The regulated power supply behaves like a constant voltage source under normal circumstances. Use this fact, together with the appropriate resistor, to create the different source resistances required.
3. Set up the source with $V_s = 10 \ V$ and $R_s = 0 \ \Omega$. Connect the values of R_L in turn and complete Table 23-2 for measured data.
4. Repeat step 3 for the two remaining values of R_s in Table 23-2, taking care to make accurate measurements of V_t and I_L for each value of R_L.
5. On the same scales and axes, plot graphs of V_t versus I_L, using the measured data from Table 23-2. Label the graphs for $R_s = 0 \ \Omega$, $R_s = 100 \ \Omega$, and $R_s = 200 \ \Omega$ with the numbers I, II, and III respectively.
6. By forming a right-angled triangle on each of the graphs, find the slope m, and together with a knowledge of V_s, write the equation of each line next to its label (see Background section).

DATA FOR EXPERIMENT 23

TABLE 23-1

Calculated Data													
	R_L (kΩ)	∞	8.2	6.2	4.7	3.6	2.7	2.0	1.5	1.1	0.82	0.62	0.47
$R_s = 0\ \Omega$	V_t												
$V_s = 10$ V	I_L												
$R_s = 100\ \Omega$	V_t												
$V_s = 10$ V	I_L												
$R_s = 200\ \Omega$	V_t												
$V_s = 10$ V	I_L												

TABLE 23-2

Measured Data													
	R_L (kΩ)	∞	8.2	6.2	4.7	3.6	2.7	2.0	1.5	1.1	0.82	0.62	0.47
$R_s = 0\ \Omega$	V_t												
$V_s = 10$ V	I_L												
$R_s = 100\ \Omega$	V_t												
$V_s = 10$ V	I_L												
$R_s = 200\ \Omega$	V_t												
$V_s = 10$ V	I_L												

NOTES

QUESTIONS FOR EXPERIMENT 23

( )

1. For $R_s = 0$, V_t should be equal to V_s regardless of load current.
   (a) True   **(b)** False

2. Refer to the graphs. For a given load current, the terminal voltage will be less for
   (a) a larger source resistance
( )
   (b) a smaller source resistance

3. The short-circuit current of the source with $R_s = 100\ \Omega$ and $V_s = 10$ V is
   (a) 10 mA   **(b)** 100 mA   **(c)** 50 mA
( )
   (d) less than that with $R_s = 200\ \Omega$

4. The slope of the line for the source with $R_s = 200\ \Omega$ is
( )
   (a) 200 Ω **(b)** $-0.005\ \Omega$ **(c)** $-200\ \Omega$ **(d)** 20 Ω

5. In light of this experiment, a "good constant voltage source" should have either a small or large source resistance. Which, and why?

6. Use your graphs to predict the approximate value of load resistance that would cause the terminal voltage of the 200 Ω source to drop to 8 V.

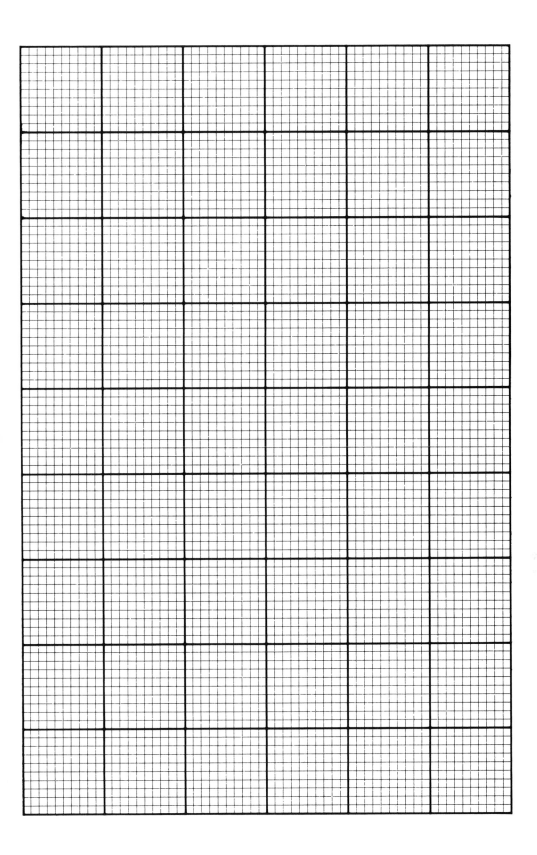

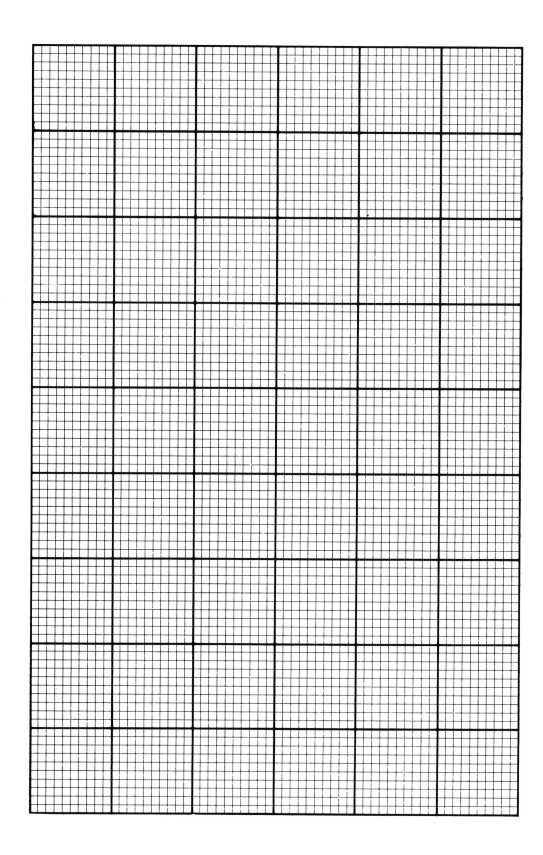

24

APPROXIMATING IDEAL VOLTAGE AND CURRENT SOURCES

REFERENCE READING

Principles of Electric Circuits: Sections 8–1 through 8–3.

RELATED PROBLEMS FROM *PRINCIPLES OF ELECTRIC CIRCUITS*

Chapter 8, Problems 1 through 6.

OBJECTIVE

To construct approximate constant voltage and constant current sources.

EQUIPMENT

dc power supply 0–20 V
DMM or VOM
Resistors (±5%): 22 Ω   10 kΩ
                       33 Ω   15 kΩ
                       47 Ω   22 kΩ
                       68 Ω   33 kΩ
                   100 Ω   47 kΩ
                  1 kΩ

BACKGROUND

In reality "perfect" or "ideal" voltage sources do not exist. All real voltage sources have a terminal voltage that will change slightly with varying load conditions. All real current sources have a load current that will change slightly with varying load conditions. The extent to which this change occurs depends upon the *relative magnitude of the effective source resistance in both cases to that of the load*. In a good voltage source, the source resistance must be much *less* than that of the load; in

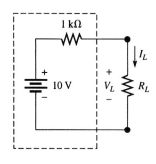

FIGURE 24-1

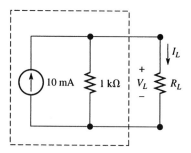

Electron flow versions of these
diagrams are in Appendix C.

FIGURE 24-2

this case the terminal voltage is approximately equal to the open-circuit value. Conversely, in a good current source, the resistance must be much *more* than that of the load to assure that the load current is approximately equal to the *short-circuit* value.

In this experiment, you will identify the terminal characteristics of approximate voltage and current sources. You will also see that the current source is nothing more than a voltage source with a very high internal resistance, compared with that of the loads we connect to it.

The circuit shown in Figure 24-1 is that of an ideal voltage source in series with some resistance that is meant to model the internal resistance of a fictitious *nonideal* source. The load voltage of this source under open-circuit conditions ($R_L = \infty \ \Omega$) is 10 V, and then, as it is loaded with different values of R_L, the load voltage changes. The object of this experiment will be to see just how close to 10 V the load voltage remains as the load resistance is changed.

The circuit in Figure 24-2 is that of a *nonideal* current source. The internal resistance of this source is 1 kΩ, and the ideal current source value is 10 mA. The load current of the source under short-circuit conditions ($R_L = 0 \Omega$) is 10 mA, and then, as the load resistance changes, the load current changes. The object of this experiment will be to see just how close to 10 mA the load current remains as the load resistance changes. Because laboratories are normally equipped with regulated (constant) voltage sources and not ideal constant current sources, some means of simulating the circuit in Figure 24-2 must be sought. The source transformation theorem allows us to transform the nonideal current source into a nonideal equivalent voltage source of the form in Figure 24-1. In fact, if you apply the equations for the *current source to voltage source transformation*, you will come up with exactly the same values as in Figure 24-1. In other words, the two circuits are identical, and whether they act like approximate voltage sources or current sources depends on the values of load resistance assigned to them.

PROCEDURE

Part A: Approximate (Nonideal) Voltage Source

1. For the circuit shown in Figure 24-1, calculate the expected values of load voltage and current for each value of lead resistance in Table 24-1.
2. The regulated power supply, together with a 1 kΩ resistor, can be used to simulate the nonideal voltage source in Figure 24-1. Set up the circuit, and measure the load current and voltage for each of the R_L values in Table 24-3. Record the measured data in Table 24-3.

Part B: Approximate (Nonideal) Current Source

1. For the circuit shown in Figure 24-2, calculate the expected values of load voltage and current for each value of load resistance in Table 24-2.

2. Since you do not have an ideal current source in the laboratory, *transform* the circuit in Figure 24-2 to an equivalent nonideal voltage source of the kind shown in Figure 24-1. The values of ideal source voltage and source resistance should calculate out to be the same as those shown in Figure 24-1.

3. The regulated power supply together with a 1 kΩ resistor can be used to simulate the nonideal source in Figure 24-2. Set up the circuit, and measure the load current and voltage for each of the R_L values. Record the data in Table 24-4.

Name _____   Date _____

DATA FOR EXPERIMENT 24

TABLE 24-1   *Calculated data*

	Approximate Constant Voltage Source				
R_L (kΩ)	10	15	22	33	47
V_L (V)					
I_L (mA)					

TABLE 24-2   *Calculated data*

	Approximate Constant Current Source				
R_L (Ω)	22	33	47	68	100
V_L (V)					
I_L (mA)					

TABLE 24-3   *Measured data*

	Approximate Constant Voltage Source				
R_L (kΩ)	10	15	22	33	47
V_L (V)					
I_L (mA)					

TABLE 24-4 *Measured data*

R_L (Ω)	Approximate Constant Current Source				
R_L (Ω)	22	33	47	68	100
V_L (V)					
I_L (mA)					

NOTES

QUESTIONS FOR EXPERIMENT 24

( )

1. A voltage source-series resistor combination will behave closely to a constant voltage source, providing that
   (a) $R_L \gg R_s$ (b) $R_L \ll R_s$ (c) $R_L = R_s$

( )

2. A voltage source-series resistor combination will behave closely to a constant current source, providing that
   (a) $R_L \ll R_s$ (b) $R_L \gg R_s$ (c) $R_L = R_s$

3. In an *approximate constant voltage source,* the load current may change dramatically as the load is changed, but the load voltage changes only a little.

( )

   (a) True   (b) False

4. In an *approximate constant current source,* the load voltage may change dramatically as the load is changed, but the load current changes only a little.

( )

   (a) True   (b) False

5. A given voltage source-series resistor combination can behave either like an approximate voltage source or an approximate current source, depending on load conditions. Explain.

6. If a nonideal voltage source has an open-circuit voltage of 20 V and an internal resistance of 20 kΩ, and if the practical rule of thumb is to be that $R_L \leq R_s/10$, over what range of load resistances will it behave like an approximate constant current source? Draw the *current source model* of this source.

25

SUPERPOSITION

REFERENCE READING

Principles of Electric Circuits: Section 8–4.

RELATED PROBLEMS FROM *PRINCIPLES OF ELECTRIC CIRCUITS*

Chapter 8, Problems 7 through 15.

OBJECTIVE

To verify the Superposition Theorem in a dc circuit.

EQUIPMENT

dc power supplies 0–12 V (two)   Resistors (± 5%): 1 kΩ
DMM or VOM   2 kΩ
                                                         3 kΩ

BACKGROUND

The superposition principle allows us to calculate the combined effects of a multi-source circuit by looking at the individual effects of each source acting alone, then summing them. Particular attention must be paid to current directions and voltage polarities when applying superposition. The experiment uses a circuit with two voltage sources, V_{s1} and V_{s2} (Figure 25-1). The three currents I_1, I_2, and I_3 can be calculated by taking the components due to V_{s1} acting alone, then V_{s2} acting alone, and adding them together. You will recall that the source not being considered is to be replaced with its internal resistance. This is the same as replacing it with a short circuit.

It is important to note that you must remove the power supply (or switch it off) before you replace it with a short circuit: never short-circuit a functioning power supply under these circumstances. Also, pay attention to the directions of the cur-

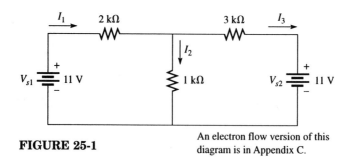

FIGURE 25-1

An electron flow version of this
diagram is in Appendix C.

rent components due to each source so that you *algebraically* sum them to get the
totals.

PROCEDURE

1. For the circuit in Figure 25-1, use the superposition principle to calculate the
   components of I_1, I_2, and I_3 due to each source acting alone, and record your
   results in the table of calculated data, Table 25-1. Note that I_{11} is the component
   of I_1 due to V_{s1}, I_{12} is the component of I_1 due to V_{s2}, etc.
2. Complete Table 25-1 by calculating the actual values of I_1, I_2, and I_3 when both
   V_{s1} and V_{s2} are active (on).
3. Connect the circuit in Figure 25-1. Remove power supply V_{s1} and replace it with
   a short circuit. Then measure the currents I_{11}, I_{21}, and I_{31}, and record them in
   the table of measured data, Table 25-2, in the appropriate column.
4. Repeat step 3, replacing V_{s1} with a short circuit. This will enable you to measure
   and record I_{12}, I_{22}, and I_{32}.
5. Finally connect the complete circuit with both sources active. Measure and record the actual currents I_1, I_2, and I_3. Compare your measured and calculated
   data.

Name _____ Date _____

DATA FOR EXPERIMENT 25

TABLE 25-1   *Calculated data*

	V_{s1} Active		V_{s2} Active		V_{s1} and V_{s2} Active (algebraic sum)
I_{11}		I_{12}		I_1	
I_{21}		I_{22}		I_2	
I_{31}		I_{32}		I_3	

TABLE 25-2   *Measured data*

	V_{s1} Active		V_{s2} Active		V_{s1} and V_{s2} Active
I_{11}		I_{12}		I_1	
I_{21}		I_{22}		I_2	
I_{31}		I_{32}		I_3	

Note: I_{11} means I_1 with V_{s1} active and V_{s2} "off."
  I_{21} means I_2 with V_{s1} active and V_{s2} "off."
  I_{31} means I_3 with V_{s1} active and V_{s2} "off."
  I_{12} means I_1 with V_{s2} active and V_{s1} "off."
  I_{22} means I_2 with V_{s2} active and V_{s1} "off."
  I_{32} means I_3 with V_{s2} active and V_{s1} "off."

NOTES

QUESTIONS FOR EXPERIMENT 25

1. With element values as in Figure 25-1, if the source V_{s1} were made equal to 22 V, then
   (a) I_3 would double
   (b) I_3 would have the same magnitude, but be opposite in direction
   (c) I_{12}, I_{22}, and I_{32} would each double

( ) (d) I_{11}, I_{21}, and I_{31} would each double

2. As far as the direction of current I_2 is concerned, the voltages V_{s1} and V_{s2}
   (a) oppose each other
   (b) aid each other

( ) (c) do neither of the above

3. If both V_{s1} and V_{s2} are increased by a factor of two, then
   (a) all currents will double
   (b) I_2 only will double
   (c) all currents will remain constant

( ) (d) I_1 and I_3 will double and I_2 will stay the same

4. The current I_3 depends on
   (a) the value of V_{s2}
   (b) the values of V_{s2}, R_3, and R_2
   (c) the values of V_{s1} and R_1

( ) (d) all of the above

5. Superposition does not work for power. For example, try using it with the power in any one of the resistors in Figure 25-1; that is, find the power in the resistor due to each source acting alone, add them, and compare with the power you get when you take the total current (or voltage) in this resistor and use the power formula. Which of these is the correct value for the power and why does superposition not work? (Hint: power depends on the square of current or voltage.)

6. With values as in Figure 25-1, calculate the value to which V_{s1} must be reduced to force the current I_1 to equal zero.

26

THEVENIN'S THEOREM

REFERENCE READING

Principles of Electric Circuits: Section 8–5.

RELATED PROBLEMS FROM *PRINCIPLES OF ELECTRIC CIRCUITS*

Chapter 8, Problems 16 and 17.

OBJECTIVE

To verify Thevenin's Theorem in a simple dc circuit.

EQUIPMENT

dc power supply 0–10 V
DMM or VOM
Resistors (±5%): 1 kΩ
                   3 kΩ (two)
                   6.2 kΩ

BACKGROUND

Thevenin's Theorem allows us to replace any circuitry (no matter how complex) "behind" two chosen terminals by a simple voltage source–series resistor equivalent. To demonstrate this we will apply the theorem to the simple three-resistor circuit in Figure 26-1. According to the theorem, we should be able to replace the circuitry to the left of terminals *A–B* in Figure 26-1 with that in Figure 26-2. To obtain the equivalent source voltage, you will measure the open-circuit voltage. The Thevenin resistance is obtained by measuring R_{A-B} with a "dead source" (replaced with a short circuit). To verify that this combination is indeed equivalent, you will then connect a load to both circuits, and verify that the resulting voltage and current are the same in both cases. Once again, do not simply short your power supply—remove it, and replace it with a piece of wire.

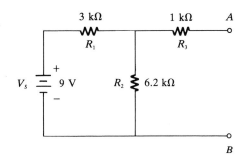

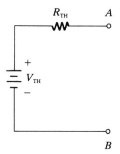

FIGURE 26-1   **FIGURE 26-2**

PROCEDURE

1. For the circuit in Figure 26-1, use Thevenin's Theorem to calculate the values of V_{TH} and R_{TH}, and record them in Table 26-1.
2. Measure the open-circuit voltage V_{AB} and record this as V_{TH} under Measured in Table 26-1.
3. Replace the source with a short circuit, and measure the resistance between the terminals A and B. Record this as R_{TH} under Measured in Table 26-1.
4. Calculate the voltage across and current through a 3 kΩ load that is to be placed across the terminals A and B. Perform the calculations for both the actual circuit and its Thevenin equivalent. The results should be identical. Record the results under Calculated in Table 26-2.
5. Connect a 3 kΩ load to the terminals A and B of the circuit in Figure 26-1. Measure the resulting load current and voltage, and record them in Table 26-2.
6. Construct the circuit of Figure 26-2 with the calculated values of V_{TH} and R_{TH}. Connect a 3 kΩ load to the terminals A and B. Measure the resulting load current and voltage, and record them in Table 26-2. They should agree closely with those in the adjacent columns.

DATA FOR EXPERIMENT 26

TABLE 26-1 *Thevenin parameters*

	Calculated		Measured
V_{TH}		V_{TH}	
R_{TH}		R_{TH}	

TABLE 26-2 *Loaded circuit parameters*

	Calculated		Measured		
			Actual Circuit		Thevenin Equivalent
I_L		I_L		I_L	
V_L		V_L		V_L	

NOTES

QUESTIONS FOR EXPERIMENT 26

( )

1. In Figure 26-1, the Thevenin resistance R_{TH} depends on
   (a) V_s (b) R_1, R_2, R_3, and V_s (c) R_1 and R_2 only
   (d) R_1, R_2, and R_3

( )

2. In Figure 26-1, the Thevenin voltage V_{TH} depends on
   (a) V_s, R_1, R_2, and R_3 (b) V_s only   (c) V_s, R_1, and R_2
   (d) V_s and R_1 only

( )

3. If R_3 were increased in value, the Thevenin voltage would
   (a) increase   (b) decrease   (c) remain the same
   (d) insufficient information

( )

4. An additional resistance placed across the terminals A and B
   (before any load is connected) would
   (a) increase R_{TH} (b) decrease R_{TH} (c) not change R_{TH}
   (d) insufficient information

5. A resistor placed directly in parallel with the source voltage V_s
   does not affect R_{TH}. Why?

6. The Thevenin equivalent resistance R_{TH} for the network in Figure 26-1 was 3 kΩ. Detail how this could be altered to 2 kΩ by using a single resistor placed across terminals A and B. Calculate the value of the resistor that will accomplish this. Will the Thevenin voltage change?

27

MAXIMUM POWER TRANSFER

REFERENCE READING

Principles of Electric Circuits: Section 8–7.

RELATED PROBLEMS FROM *PRINCIPLES OF ELECTRIC CIRCUITS*

Chapter 8, Problems 30 through 33.

OBJECTIVE

To verify the Maximum Power Transfer Theorem for dc circuits.

EQUIPMENT

dc power supply 0–10 V
DMM or VOM
Resistors ($\pm 5\%$):

1 kΩ	3.3 kΩ
1.5 kΩ	4.3 kΩ
2 kΩ	5.6 kΩ
2.7 kΩ	7.5 kΩ
3 kΩ (two)	10 kΩ

BACKGROUND

In this experiment you are going to answer the question, "Given a voltage source with some fixed internal resistance R_s, what value of load resistor will absorb the most power from the circuit?" Clearly, there are an infinite number of loads from 0 Ω (a short circuit) to ∞ Ω (an open circuit). Since a short circuit would have no voltage and an open circuit no current, both of these produce zero load power. The answer lies somewhere between 0 Ω and ∞ Ω. The maximum power transfer theorem tells us that the load resistance should be equal to the source resistance for maximum power to be absorbed by the load.

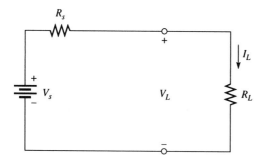

FIGURE 27-1

In the experiment you will connect a series of increasing load resistors to a source resistor combination. In each case, you will measure V_L and I_L and calculate P_L $(= I_L V_L)$, the power in the load. You should find that this power is maximized where $R_L = R_s$. A graph is one of the best ways to illustrate the theorem. Be sure to join your data points with a smooth curve (not straight-line segments) so that the true variation of power with load resistance can be observed.

You will also plot the circuit's efficiency versus the load resistance. The efficiency measures just what fraction of the total power dissipated is dissipated in the load resistance. That is,

efficiency $= P_L/P_T$

where P_L is the load power and P_T is the total power (some power is always dissipated in the source resistance). You will see that when delivering maximum power to the load, the efficiency is *not* a maximum.

PROCEDURE

1. For a value of V_s equal to 10 V, and $R_s = 3$ kΩ, calculate the load voltage V_L, load current I_L, and load power P_L in Figure 27-1 for each value of R_L in Table 27-1.
2. Complete the table by calculating the total power ($P_T = V_s \times I_L$) and the circuit's efficiency ($P_L/P_T \times 100\%$) for each value of load resistance.
3. Set up the circuit and, for each value of R_L, measure the load current and voltage. Record all measured data in Table 27-2. From these data, complete the table by calculating the total power and efficiency. The measured data should compare reasonably well with theory.
4. Plot a graph of the load power versus R_L from your *measured* data. On *different* scales and axes, plot the circuit efficiency, also from your measured data.

Name _____ Date _____

DATA FOR EXPERIMENT 27

TABLE 27-1 *Calculated data*

R_L (kΩ)	1	1.5	2	2.7	3	3.3	4.3	5.6	7.5	10
V_L (V)										
I_L (mA)										
P_L (mW)										
P_T (mW)										
Efficiency (%)										

TABLE 27-2 *Measured data*

R_L (kΩ)	1	1.5	2	2.7	3	3.3	4.3	5.6	7.5	10
V_L (V)										
I_L (mA)										
P_L (mW)										
P_T (mW)										
Efficiency (%)										

NOTES

QUESTIONS FOR EXPERIMENT 27

( )
1. The load power is maximum when
   (a) $R_L \gg R_s$ (b) $R_L \ll R_s$ (c) $R_L = R_s$
2. If the source resistance were made larger than 3 kΩ in this experiment, then
   (a) all the values of P_L would be less
   (b) all the values of P_L would be greater
   (c) the values of P_L would not change

( )
   (d) some values of P_L would be smaller and some greater

( )
3. The power dissipated in the source resistance is always
   (a) $\leqslant P_L$ (b) $< P_L$ (c) $\geqslant P_L$ (d) none of these
4. The power dissipated in the source resistance depends on the value of the load resistance.

( )
   (a) True   (b) False
5. To achieve high levels of efficiency, what is the required relationship between the source and load resistance?

6. Show that the efficiency of the source is only 50 percent when supplying its maximum power.

Efficiency (%)

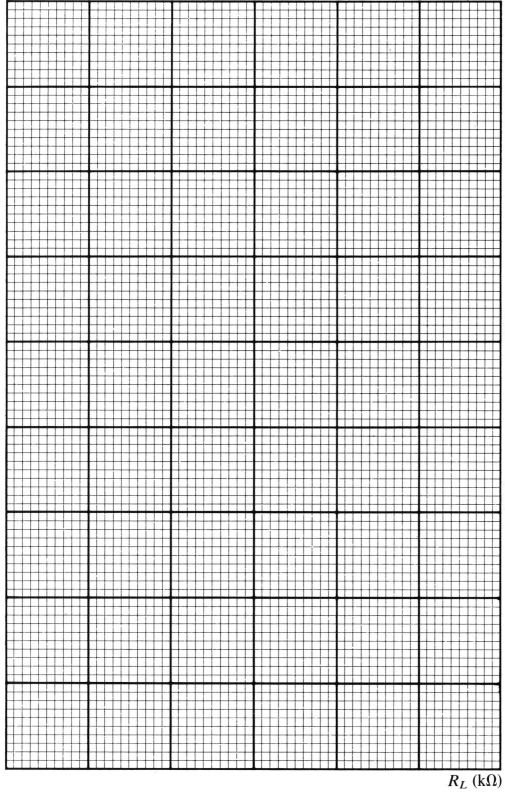

R_L (kΩ)

P_L (mW)

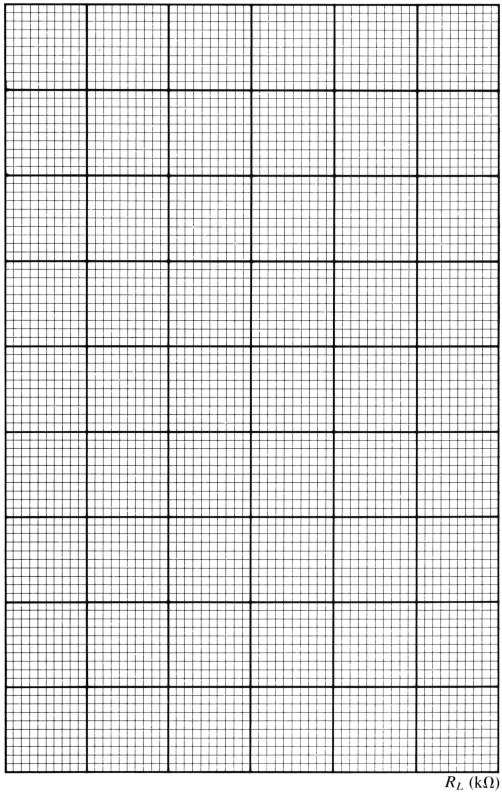

R_L (kΩ)

28

BRANCH, LOOP, AND NODE ANALYSES

REFERENCE READING

Principles of Electric Circuits: Sections 9–1 through 9–5.

RELATED PROBLEMS FROM *PRINCIPLES OF ELECTRIC CIRCUITS*

Chapter 9, Problems 17 through 31.

OBJECTIVE

To be able to write mesh and node equations for a simple resistive circuit, and to prove, through measurement, that these methods work.

EQUIPMENT

dc power supply 0–12 V
DMM or VOM
Resistors (±5%) 2 kΩ
              3 kΩ
              1 kΩ

BACKGROUND

The methods known as loop analysis and nodal analysis are generalized circuit analysis techniques that are particularly useful in the analysis of circuits with multiple loops and at least two sources, and where the methods of ordinary series–parallel resistive circuits fail. The methods use Ohm's law together with one of the

Kirchhoff laws. In both cases a linear system of equations is produced, which must be solved. Floyd shows how these systems can be solved using Cramer's Rule or a linear system solver such as is built into many graphing calculators.

Part A: Loop Current Method

In the loop current method as explained in the text, we assign (arbitrarily) loop currents around each loop of the circuit. For example, in the circuit of Figure 28-1, we have assigned the two loop currents, I_1 and I_2, each in the clockwise direction. We now set up the system of loop equations as follows. All resistances are in $k\Omega$.

Going around loop 1, with Kirchhoff's Voltage Law we have,

$$2I_1 + 1(I_1 - I_2) = 11 \qquad \text{for loop 1}$$

while loop 2 gives

$$3I_2 + 1(I_2 - I_1) = -11 \qquad \text{for loop 2}$$

After rearranging the equations in standard form, we get

$$(2 + 1)I_1 - 1I_2 = 11 \qquad \text{for loop 1}$$
$$-1I_1 + (3 + 1)I_2 = -11 \qquad \text{for loop 2}$$

When these equations are solved using either of the methods described in the text, we get

$$I_1 = 3 \text{ mA and } I_2 = -2 \text{ mA}$$

The negative sign on I_2 tells us that the direction we assigned for this current is opposite of the actual current direction.

The resistor currents labeled I_{R1}, I_{R2}, and I_{R3} on the same diagram can now be computed as follows.

$$I_{R1} = I_1 = 3 \text{ mA}$$
$$I_{R2} = I_1 - I_2 = 3 - (-2) = 5 \text{ mA}$$
$$I_{R3} = -I_2 = 2 \text{ mA}$$

Part B: Node Voltage Method

Refer to Figure 28-2 in which the circuit of Figure 28-1 has been relabeled to reflect the pertinent variables for nodal analysis. In the node-voltage method, we select one node (frequently circuit ground) as the reference node. In this case we will assign the node at the bottom of the circuit (node B) as the reference node. There are three remaining nodes, A, C, and D. The voltages at nodes C and D are known since they are given by the source voltages. The only unknown node voltage is at node A. We

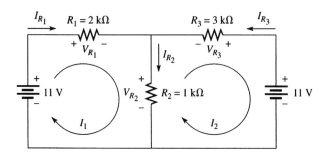

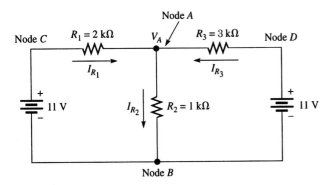

FIGURE 28-2

now use Kirchhoff's Current Law to sum the currents at node A. Using the branch (resistor) currents, I_{R1}, I_{R2}, and I_{R3}, we can write

$$I_{R1} - I_{R2} + I_{R3} = 0$$

Now, expressing the currents in terms of the circuit voltages, we have

$$I_{R1} = \frac{11 - V_A}{2}$$

$$I_{R2} = \frac{V_A}{1}$$

$$I_{R3} = \frac{11 - V_A}{3}$$

and substituting these into the current equation gives us

$$\frac{11 - V_A}{2} - \frac{V_A}{1} + \frac{11 - V_A}{3} = 0$$

which yields

$$V_A = 5 \text{ V}$$

from which the remaining circuit voltages and currents can be found.

PROCEDURE

Part A: Mesh Analysis

1. Measure and record the values of the resistors and record the data in Table 28-1.
2. Connect the circuit in Figure 28-1. Use the method of loop equations as shown in the Background section to determine the loop currents I_1, I_2 and the resistor currents I_{R1}, I_{R2}, and I_{R3}. Record this data in Table 28-2.
3. Using the computed currents in each resistor, calculate the expected voltage drops across each of the three resistors. Record this data in Table 28-3.
4. Measure the actual voltages across each resistor and record them in Table 28-3 also, for comparison.

Part B: Node Analysis

1. Write down the node equation at node A for the circuit in Figure 28-2. Use the measured resistance values from step 1 of Part A of the experiment.

2. Solve the node equation for node voltage V_A. Record this node voltage in Table 28-4.
3. Using the measured resistor values, compute the resistor currents I_{R1}, I_{R2}, and I_{R3}. These values can be recorded in Table 28-4. The voltage V_A as well as the calculated currents should compare well with those in the previous part of the experiment.

DATA FOR EXPERIMENT 28

TABLE 28-1

Resistor	Nominal	Measured
R_1	2 kΩ	
R_2	1 kΩ	
R_3	3 kΩ	

TABLE 28-2

	Computed Currents
I_1	
I_2	
I_{R1}	
I_{R2}	
I_{R3}	

TABLE 28-3

Voltages	Computed	Measured
V_{R1}		
V_{R2}		
V_{R3}		

TABLE 28-4

	Computed Quantities
V_A	
I_{R1}	
I_{R2}	
I_{R3}	

QUESTIONS FOR EXPERIMENT 28

1. Suppose that the loop current I_2 in Figure 28-1 had been defined counterclockwise instead of clockwise, and a new set of loop equations had been written, what would have been the effect on the value of I_2 in the solution of this new system of loop equations? What would have been the effect on the value of I_2 in the solution of the loop equations?

(   )   (a) I_2 would still equal 2 mA   (b) I_2 would equal -2 mA

2. With loop current I_2 defined as in the previous question, then

(   )   (a) $I_{R2} = I_1 - I_2$ (b) $I_{R2} = I_1 + I_2$ (c) $I_{R2} = I_2 - I_1$

3. Nodal analysis uses Ohm's law together with

(   )   (a) Kirchhoff's Voltage Law   (b) Kirchhoff's Current Law

4. Suppose that we had chosen V_A as the reference voltage for nodal analysis. Then the node voltage at node B would have been

   (a) -11 V   (b) -22 V   (c) -5 V   (d) 5 V

(   )   (e) -16 V

5. Recalculate the circuit voltages and currents using loop analysis if the source on the right was adjusted to 22 V.

6. Repeat the previous question but use node analysis.

II

ac EXPERIMENTS

A NOTE ON ac EXPERIMENTS

For the beginning student, ac experiments often do not give the nice results that one obtains with those in dc. This is sometimes due to poorly designed experiments, but more often, it is because certain basic precautions have not been taken. One of the most serious problems that the beginning student faces is an inadequate knowledge of the test instruments, particularly the oscilloscope and function generator. The early experiments in Part II are designed to acquaint you with these basics so that, in later experiments, you can concentrate on the nature of the activities rather than on how the equipment is to be used. What follows are some guidelines that, if put into practice, will help you avoid some of the more common pitfalls that can occur in ac work.

Circuit Ground

For safety and other reasons, one side (input or output terminal) of the electrical system and the chassis (case) of almost all signal generators and oscilloscopes are *earth-grounded* through the third pin of the plug that supplies ac power to the equipment. In the case of signal generators, a chassis ground marking ($\overline{7\!\!\!/\!\!7\!\!\!/\!\!7}$) or **LOW** designation might appear next to the grounded terminal. In any case, this usually means that this terminal is literally connected to the earth's surface. It also means that there is a hidden connection between all instruments that are grounded in this manner. When working with a circuit in which both instruments are connected at one time, one must be careful to ensure that the grounded leads are *never* connected to electrically different points in the circuit. If two grounded leads from two earth-grounded instruments are connected to different points in the circuit, then there will be an unintended short circuit between these two points which will inevitably result in bad data and can sometimes produce an electrically dangerous condition. Your instructor should properly explain this concept to you. Beginning students have a particularly difficult time with the concept.

Proper Use of Meters

Just as in dc work, in ac work we can use either digital or analog meters in addition to the oscilloscope. You will see that the *dc* ranges on either kind of meter can be used to give meaningful information about ac waveforms. This usually presents no problem as long as the frequencies are not too low (less than about 20 Hz). However, one must be very careful when using these meters on their *ac* ranges. It is here that you may run into trouble because all meters have a useful operating frequency range beyond which the accuracy of the readings cannot be relied upon. Students who are unaware of this limitation will attempt to use inexpensive meters at high frequencies for which they were not designed. Always make sure that the meter you are using can operate properly at the frequencies appearing in the circuitry. Operations manuals or your instructor can inform you of this. Be aware that many ordinary (and inexpensive) general-purpose meters can operate up to only about 500 Hz, after which their indications cannot be relied upon.

The Oscilloscope

Oscilloscopes are probably the most useful of all general-purpose test instruments and are also one of the more complex that you will use in your class. There are two experiments in this section designed to get you off to a good start at understanding the operating features of a standard dual-trace oscilloscope. Keep in mind that you cannot hope to feel entirely comfortable with this instrument after two experiments and only a few hours of time. As you progress through the ac experiments, however, you will become increasingly confident with the instrument, and by the end of the book, you will have used all but a few special controls on the oscilloscope front panel.

ac Loading—Measurement Errors

Just as with dc circuits, instruments always have some effect on the circuitry to which they are connected and can cause large discrepancies between theoretical and apparent measured values. The loading problem is especially troublesome in ac because it tends to worsen at high frequencies. In circuits that are operating at higher frequencies or if they have high impedances, your instructor should have you use $10 \times$ probes with the oscilloscope. These special probes reduce the loading effect that the 'scope has on the device under test. Before you use them, they must be *frequency compensated,* which is a simple process you can do yourself. Your instructor should show you how this is accomplished. An uncompensated $10 \times$ probe is as useless as not having one at all, so always compensate the probe before you use it for measurements.

The Function (Signal) Generator

Another misconception students have is that the voltage at the output terminals of function/signal generators remains constant once set. These generators are not like the dc power supplies we used in our dc work. Those were regulated sources. Generators of ac waveforms have a substantial but well-defined internal, or output, resistance, that causes the voltage across the terminals of the generator to be dependent on the load. Many of the experiments in Part II require a constant-amplitude signal at the load over a wide frequency range. You should always expect the terminal voltage of the generator to change with frequency because the load generally changes with frequency. The only way to maintain this constant-amplitude output is for you to keep a watchful eye on the voltage and change it when necessary.

Components in ac Circuits

Finally, remember that the capacitors and inductors that you will be using will have broader tolerances than the 5 percent resistors with which you have become familiar. They also have imperfections that may present problems not normally encountered in dc experiments. For example, because of phenomena known as skin effect and hysteresis and for other reasons, inductors have an ac resistance that is not apparent from a simple ohmmeter test. They also have a parasitic (stray) capacitance, which can have odd effects at higher frequencies. Capacitors have lead inductance, which should not be a problem in these experiments, but the imperfect dielectric of the capacitor might. If, during an experiment, your measurements seem to be substantially different from what you expected, see your instructor for help. It is otherwise very frustrating for the student who goes away and attempts to process his or her results only to find that they have no resemblance to the expected values.

29

OSCILLOSCOPE FAMILIARITY, I

REFERENCE READING

Principles of Electric Circuits: Section 11–10.

RELATED PROBLEMS FROM *PRINCIPLES OF ELECTRIC CIRCUITS*

None.

OBJECTIVE

To become familiar with the basic operating controls of the dual-trace oscilloscope. To learn how to use the oscilloscope to measure dc voltages.

EQUIPMENT

Dual-trace oscilloscope and $1 \times$ probe
dc power supply 0–10 V
DMM

BACKGROUND

The cathode ray oscilloscope is one of the most flexible and useful of the instruments available to the technician. Though it is most often used to display a time-varying voltage, it can also be used in a manner similar to that of a dc voltmeter. The purpose of this experiment is to familiarize you with the basic operating controls and then to use the instrument in the determination of dc voltage. To beginning students of electronics, the instrument is, at first, formidable, but after a few sessions with it, you will begin to feel as comfortable with it as you do with a meter.

If you flip through the pages of this experiment, it looks rather lengthy. In fact, much space is devoted to teaching you how to do an initial (default) setup; then how some of the controls affect measurement is explained; and finally you will use the oscilloscope to measure some dc voltages.

It is important that you go slowly and thoroughly through this particular experiment because the oscilloscope will be used in almost every experiment in the ac section of the manual. If it helps, take notes, so that you are able to remember important results.

Try not to be intimidated by the instrument, and spend as much time as you need to on each of the controls. By all means, at the end of the activity, if time permits, "play" with each of the operating controls and develop some confidence in your ability to use this versatile instrument. There is some excellent material in the PEC text on using the oscilloscope. Read it before you come to class to do this experiment.

PROCEDURE

For emphasis, oscilloscope operating controls and labels appear in boldface uppercase type.

Oscilloscope Connections and Grounding

1. The oscilloscope is an *earth-grounded* instrument; that is, one side of the electrical system of the instrument is connected to the earth via the line cord that connects it to the electrical supply. This can be verified using the DMM on **OHMS.** Using the DMM, set to **OHMS,** check to see which parts of the oscilloscope are connected to the ground pin of the line plug. The chassis and the outer shield of all the Bayonet Neill-Concelman* (BNC) input connectors should be connected to earth ground. Also, if there is a specific terminal marked with the earth ground symbol ($\perp$), it, too, should have continuity with the line plug ground pin.

2. Because each of the outer conductors on the input connectors are connected to earth ground, this means that *each is connected to the other, and therefore the ground leads of any probes connected to these inputs must always be connected to points at the same electrical potential in the circuit under test.* The beginning student often fails to realize this, and as a result, many inadvertent short circuits are produced, resulting in incorrect measurements and/or electrically hazardous conditions.

3. Examine one of the oscilloscope probes. The end that connects to the oscilloscope has a special device called a BNC connector attached to it. The outer "shell" of the connector is the grounded side because it is connected to the earth ground of the oscilloscope measuring system. The inner part of the connector is sometimes referred to as the signal side.

   At the other end of the lead, both the ground and signal leads are brought out, so that they can be attached to the observation points in your circuit. This end of the cable is sometimes referred to as the *sampling end,* and there is often a black alligator clip-lead, which is internally connected to the grounded side of the oscilloscope.

Default Control Settings

4. For a beginning electronics student, it represents good practice to set up the various controls of an oscilloscope in a standard way. These standard positions for the controls might be called "default settings." They can be done before or

---

*This connector was developed by Paul Neill and Carl Concelman at Bell Laboratories.

after the oscilloscope has been switched on. The default settings assume that the user intends to use the oscilloscope in the *dual-trace sweep mode;* that is, to display two waveforms simultaneously, one on each input channel.

The following instructions work particularly well with oscilloscopes manufactured by Tektronix Inc. However, you should be aware that various oscilloscopes may have slightly different labels for the same controls and that some oscilloscopes may not have some of them at all. If in doubt, see your lab instructor. The oscilloscope panel has been divided into four main functional blocks for this purpose; these are the display section, the vertical section, the horizontal section, and the trigger section.

5. *The display section*
   (a) **INTENSITY**—turn fully counterclockwise, then turn about halfway up.
   (b) **FOCUS**—turn fully counterclockwise, then turn halfway up.
   (c) **BEAM-FINDER**—leave as is.
6. *The vertical section*
   (a) **VERTICAL MODE** switches—set to **BOTH** and **CHOP.**
   (b) **Ch. 1** and **Ch. 2 POSITION** controls—turn fully counterclockwise, then turn about halfway up.
   (c) **Ch. 1** and **Ch. 2** vertical attenuator (**VOLTS/DIV,** or **V/DIV**) settings—1 **VOLT/DIV** as viewed through the $1\times$ window. (Some oscilloscopes have both a $10\times$ and a $1\times$ viewing window on this control.)
   (d) **Ch. 1** and **Ch. 2** vertical attenuator **VARIABLE** (vernier) controls—in **CAL**ibrate position (usually fully clockwise—turn in the direction of the arrow).
   (e) **Ch. 1** and **Ch. 2** input coupling switches—**DC.**
   (f) **INVERT** (usually provided for channel 2) in the **OUT** (noninvert) position.
7. *The horizontal section*
   (a) Horizontal **POSITION** control—turn fully counterclockwise, then turn about halfway up.
   (b) **HORIZONTAL MODE** control—set to **NO DELAY.**
   (c) Horizontal attenuator **(SECS/DIV)** settings—1 **ms/DIV.**
   (d) Horizontal attenuator **VARIABLE** (vernier) controls—in **CAL**ibrate position (usually fully clockwise—turn in the direction of the arrow).
   (e) **DELAY TIME** (if provided)—setting is not critical.
   (f) **DELAY TIME MULTIPLIER** (if provided)—setting is not critical.
8. *The trigger section*
   (a) **VAR HOLDOFF—NORMAL** position.
   (b) **TRIGGER MODE** switch—in **AUTO** position.
   (c) **TRIGGER SLOPE** switch—rising edge ($\underline{\mathcal{J}}$) or positive slope setting.
   (d) **TRIGGER LEVEL** control—turn fully counterclockwise, then turn about halfway up.
   (e) **TRIGGER SOURCE** switch—on **INT.**
   (f) **INT TRIGGER** selection—set to **Ch. 1.**
   (g) **EXT COUPLING** switch—setting not critical.
9. If the oscilloscope **POWER ON** switch is still in the **OFF** position, turn on the oscilloscope. The default settings above will normally result in a display of two distinct horizontal lines. These lines will be of random intensity, focus, and position. If the lines are still not visible, use the **BEAMFINDER, POSITION, INTENSITY,** and **FOCUS** controls to bring them into view.

This completes the default settings for the dual-trace oscilloscope. Try to commit them to memory, so that you are able to perform them without thinking each time you begin to use the oscilloscope. *Note:* Some of the controls generally require further adjustment depending on how you intend to use the oscilloscope once the waveforms to be observed are connected to the inputs.

Setting Up a Ground Reference Trace

10. The horizontal lines you have displayed are randomly positioned on the screen. Under normal circumstances, a *zero voltage reference line* is set up. To do this, proceed as follows:
    (a) Set both the **Ch. 1** and **Ch. 2** input coupling switches to **GROUND.** This electrically isolates any voltage waveform at the sampling ends of the probe *and* forces the voltage at the inputs to the oscilloscope to 0 V.
    (b) Use the **POSITION** control associated with **Ch. 1** to position the beam so that it coincides with the *horizontal line* one-quarter of the way down from the top of the graticule.
    (c) Use the **POSITION** control associated with **Ch. 2** to position the beam so that it coincides with the *horizontal line* one-quarter of the way up from the bottom of the graticule. The traces should look as shown in Figure 27-1.
    (d) Restore the input coupling switches to **DC.**

Initial Check-out and Functional Test

11. On the front panel of most oscilloscopes, there is normally a terminal or jack with the words **probe adj** or **CAL** inscribed. At this terminal, an internally generated square wave of fixed amplitude and frequency is available. Its main purpose is to provide a signal for the adjustment of the various kinds of probes that can be used in conjunction with the oscilloscope. However, it also provides us with a convenient way to determine if the oscilloscope's basic operational controls are in good working order. Locate this output jack. Connect a probe to the **Ch. 1** input, and hook up the sampling end of this output jack. The input coupling switch for channel 1 should be set to **DC,** and the **TRIGGER MODE** switch to **AUTO** (the default settings).

12. A low-amplitude rectangular wave should be displayed on the screen. If the signal is not visible, or appears too small, adjust the vertical attenuator **(VOLTS/DIV)** switch to a *more sensitive* setting, such as 0.5 **VOLTS/DIV.**

13. Turn the vertical attenuator switch to different settings. The p–p deflection on the screen should get *larger* as you switch to *more sensitive* settings. (*Note: More* sensitive equals *fewer* **VOLTS/DIV.**)

14. Likewise, turn the horizontal attenuator **(SECS/DIV)** switch to different settings. The signal should "spread out" in the *X* direction (which represents time) as you switch to more sensitive settings (higher sweep speeds).

15. Adjust the **INTENSITY** and **FOCUS** controls so that the appearance of the trace is to your liking.

The preceding procedure is a quick, convenient way to determine the functional integrity of some of the basic operating controls of the oscilloscope. *It does not, however, prove that other functions, such as triggering, etc., are working adequately, nor does it show whether the oscilloscope is correctly calibrated.*

The Oscilloscope as a dc Voltmeter

Though the oscilloscope is capable of much more sophisticated measurements, its use in measuring the magnitude and polarity of simple dc voltages is common and deserves some attention.

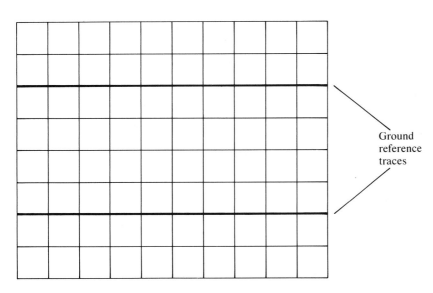

Ground
reference
traces

FIGURE 29-1

16. Set up the vertical mode switches to view only **Ch. 1.** Reposition the ground reference trace so that it lies in the center of the graticule. Be sure to set the input coupling switch for **Ch. 1** to **GROUND** when you do this.

17. Switch on the dc power supply, and use the panel meter or a DMM to set its output to some random value of voltage between 1 and 4 V.

18. With the **Ch. 1** attenuator at 5 **VOLTS/DIV,** connect the output of the dc power supply to the **Ch. 1** input, so that the negative terminal of the supply is connected to the oscilloscope ground (normally a black alligator lead) and the positive terminal to the probe tip. With input coupling for **Ch. 1** on **DC,** you ought to be able to see a steady trace on the screen.

19. Gradually turn up the dc voltage at the power supply, and watch the trace on the screen. The horizontal line that you see indicates the following:

   (a) The magnitude of the voltage. This is obtained by multiplying the vertical **VOLTS/DIV** setting (symbolized as V below) by the number of major divisions Def (in centimeters), from the ground reference position:

$$V_{dc} = V \text{ (\textbf{VOLTS/DIV})} \times Def \text{ (\textbf{DIV})}$$

   (b) The polarity (with respect to *earth ground*) of the voltage.

   (c) A constant voltage (not changing with time).

   Using the best **VOLTS/DIV** setting in each case, use the oscilloscope to measure some randomly chosen dc voltages and record the data in Table 29-1. Confirm these voltages with your DMM.

20. With a setting of 1 **VOLT/DIV,** and a centered ground reference trace, you can measure voltages up to a maximum of $+4$ V. By changing the vertical attenuator setting to its *least sensitive* position and moving the ground reference trace to the lowermost line of the graticule, what is the largest magnitude voltage you are able to reliably display under these conditions (without the trace disappearing off the top of the screen)?

21. Switch off the supply. Check to see whether either of the power supply terminals is *earth-grounded* via the ground pin on its own line plug. Use the DMM on **OHMS** for this. If neither of the power supply terminals is earth-grounded, then the power supply is said to be floating, and it is permissible to connect

either terminal to an *earth ground* through the oscilloscope. Switch* the probe connections to the power supply so that the *positive* side of the supply is connected to the *grounded* side of the oscilloscope probe and the negative side to the probe tip.

22. Starting with a centered ground reference line, once again slowly increase the power supply voltage. This time the trace should move downward toward the bottom of the screen. The polarity of the dc voltage with respect to earth ground is now negative. The magnitude and polarity of the voltage can be calculated as described, by calling the deflection a *negative* one.

Effect of Input Coupling Switch on dc Voltage Measurement

23. Set up the power supply with a positive dc voltage represented on the oscilloscope screen. Now flip the **Ch. 1** input coupling switch to **GROUND.** *Did you notice that the trace returns to its ground reference position even though you have a dc voltage on the input?* This is because this switch *disconnects* the input voltage from the oscilloscope input while at the same time replacing it with a 0 V (ground) condition at the oscilloscope input. Note that the power supply is *not shorted* out because the switch performs a disconnect function. This means that when you select the **GROUND** position with the input coupling switch, you do not have to physically disconnect the probe tip from the device under test.

24. Restore the input coupling to **DC.** The dc voltage reappears. Now flip the switch to **AC.** The trace returns to the grounded position. Although it may appear that the effect on the trace is the same as that of the **GROUND** position, it is not. In this position, a capacitor is inserted in series with the input signal, and so the dc voltage is blocked. The appearance of a 0 V level trace on the screen is to be expected. Use of the **AC** position is best examined when we begin to look at ac signals with dc components and when you understand the use of capacitors at a later point. From this, you can conclude that when observing dc voltages the *input coupling switches* must be set to the **DC** position.

Effect of Trigger Mode Switch on dc Voltage Measurement

25. Set the output to some dc voltage so that you have a measurable deflection on the screen. Locate the **TRIGGER MODE** switch. Flip it to **NORMAL,** then back again to **AUTO,** watching the trace as you do so. Did you notice that the trace is visible only under the **AUTO** trigger mode? The **AUTO** trigger mode is the *only* position that you can be *certain* will result in a visible trace in the absence of a *changing* signal at the vertical inputs to the oscilloscope. This is why we choose **AUTO** as the default position for this switch. (The **TV** setting on some oscilloscopes will also display a trace with no changing input signal, but **AUTO** is the best default choice.) Since *you have* a dc input signal (that is, a signal that is not changing), then the **AUTO** trigger mode is the only setting that allows you to see the trace. *In fact, unless the input signal is changing, the **AUTO** trigger mode should always be used to get the trace to appear on the screen.*

---

*WARNING: The power supply must have floating terminals if you are to do this. If you are in any doubt, ask your instructor before proceeding; this procedure can damage some power supplies.

Name _____ Date _____

DATA FOR EXPERIMENT 29

TABLE 29-1

Deflection (DIV)	Vertical Attenuator (VOLTS/DIV)	Voltage

NOTES

QUESTIONS FOR EXPERIMENT 29

1. When setting up an oscilloscope, the best default setting for the **TRIGGER MODE** switch(es) is

( )   **(a) NORMAL**　　**(b) TV**　　**(c) AUTO**　　**(d)** any of these

2. Horizontal and vertical **VAR**iable controls are normally
   (a) left in whatever condition they are in when you switch on
   (b) set in their calibrated **(CAL)** position

( )   **(c)** set halfway up from fully counterclockwise

3. When setting up a ground reference trace, the *input coupling switches*
   (a) should be set to **DC**
   (b) should be set to **AC**
   (c) should be set to **GROUND**

( )   **(d)** can be left in any position (it does not matter)

4. Which of the following horizontal sensitivity **(SECS/DIV)** settings are *not* available on a modern oscilloscope?
   (a) 1 ms/DIV　　**(b)** 2 ms/DIV　　**(c)** 4 ms/DIV

( )   **(d)** 5 ms/DIV

5. Suppose you are monitoring a dc voltage, and you set the *input coupling switch* to **AC**. What will you see and why?

6. Why is it important to know if some parts of a circuit are connected to *earth ground* when you are using an oscilloscope to view voltages on this circuit?

30

OSCILLOSCOPE FAMILIARITY, II

REFERENCE READING

Principles of Electric Circuits: Section 11–9 through 11–10.

RELATED PROBLEMS FROM *PRINCIPLES OF ELECTRIC CIRCUITS*

Chapter 11, Problems 43 and 44.

OBJECTIVE

To become familiar with the basic operating controls of the dual-trace oscilloscope. To learn how to use the oscilloscope to measure the peak-to-peak amplitude and period of ac voltages.

EQUIPMENT

Dual-trace oscilloscope and 1× probes
Audio function generator with square, triangle, and sine-wave outputs
DMM

BACKGROUND

In experiment 29, you became familiar with the basic operating controls of the oscilloscope. You should now be able to get a trace to appear by following the default settings that were outlines in that experiment.

In this experiment we will look at how the oscilloscope can be used to display and measure the characteristics of ac waveforms—voltages that vary with time. In particular, we will look at using the oscilloscope in the determination of the amplitude and period of some simple waveforms that you can get from a function generator. Perhaps the most important default settings from experiment 29 are the selection of **AUTO** for the trigger mode, and **DC** for the input coupling switches. You should also make sure that the **VAR**iable **VOLTS/DIV** and **SECS/DIV** controls are in their **CAL**ibrated positions.

Though it is not until experiments 31 and 32 that you will become completely familiar with the function generator, we will have to use this instrument in this activity. However, you will use only a few of its controls.

PROCEDURE

1. Initialize all oscilloscope controls to their default settings, as explained in experiment 29. Adjusting the oscilloscope to view the channel 1 signal only, set up a ground reference trace for channel 1 in the center of the graticule, and make sure that all variable controls are in their **CAL** positions.

2. Turn your attention to the function generator. Locate the function generator's output terminals. Locate and identify the generator's frequency control/switches and also its amplitude (ac output) control.

   Note that some function generators have an *unbalanced output;* that is, one of the terminals might be *grounded,* and it is therefore important, when you proceed to step 3, that you connect this **LOW** side of the generator to the *grounded* side of the oscilloscope. See your instructor if in doubt about this.

3. With its output connected to the **Ch. 1** input of the oscilloscope, set up the function generator for a 500 Hz *symmetrical* (zero the dc offset control on the generator if it has one) *triangle wave* output, as shown in Figure 30-1 (going both positive and negative with respect to ground). If the generator has no triangle output, use a sine wave instead. Note that you need only adjust the generator's amplitude control until the waveform takes up four major divisions in the vertical direction. If the waveform does not appear stationary, rotate the trigger **LEVEL** control slightly until it does. (We will be looking at the level control in some detail at a later point.) The waveform that you see might not "start" at the same point as it does in the figure; however, do not worry about this at this juncture.

4. Determine the p–p value of the waveform by measuring the distance in major divisions (one major division is equal to one centimeter) from the uppermost to the lowermost peaks of the displayed waveform. Then proceed with the calculation as follows:

$$V_{p-p} = V\,(\textbf{VOLTS/DIV}) \times Def\,(\textbf{DIV})$$

where V is the vertical attenuator setting and Def is the number of major divisions from the *positive to the negative peak of the waveform.* When you calculate the p–p value of the waveform, you should get an answer of 4 V p–p.

5. Turn the channel 1 variable (vernier) control *out* of its **CAL**ibrated position. The displayed signal will be reduced in amplitude. You should be aware that the *actual signal level* at the function generator output has not changed—only the scale factor by which it is displayed on the oscilloscope screen. This control allows you to make comparative measurements between signals when you are *not* specifically interested in the actual signal level. *Restore the control to its* **CAL** *position.*

   It should be clear that you have no way of knowing the actual p–p value of a signal displayed when the vernier is not in its **CAL** *position.*

6. Expand the **SECS/DIV** control to 0.5 **ms/DIV.** You should have about two and one-half cycles of the triangle waveform displayed on the screen. You should realize that the actual signal frequency or period has not changed, but the horizontal scale factor has been halved to 0.5 **ms/DIV.** Determine the period of the waveform by measuring the distance in major divisions (one major di-

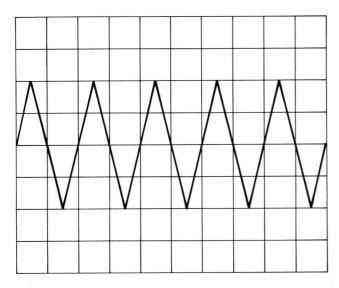

FIGURE 30-1

vision is equal to one cm) from one zero crossing to the next *corresponding* zero crossing of the displayed waveform. Use the horizontal **POSITION** control to adjust the position of the waveform to your liking. Then proceed with the calculation as follows:

$$T \text{ (secs)} = H \text{ (\textbf{SECS/DIV})} \times Def \text{ (\textbf{DIV})}$$

where H is the horizontal timebase setting, and Def is the number of major divisions (centimeters) between corresponding zero crossings. When you calculate the period of the waveform, you should get an answer of 2 ms. You should be able to confirm the frequency by taking the reciprocal of this period (i.e., $f = 1/T$).

7. Restore the **SECS/DIV** to 1 **ms/DIV.** Switch the signal to a square wave and then a sine wave, and compare the traces. Each waveform ought to have the same frequency and p–p value. Only the shape of the waveform should look different.

8. Restore the generator setting to the original waveform. Turn the **SECS/DIV** control one position counterclockwise. The display will have been compressed so that you see one cycle of the waveform contained within each major division on the graticule. You should realize that the actual signal frequency or period has not changed, but the horizontal scale factor has been doubled to 2 **ms/DIV.**

9. Restore the horizontal attenuator setting to 1 **ms/DIV.** Turn the horizontal variable (vernier) control out of its **CAL**ibrated position. The display will be compressed in the horizontal direction. Once again, you should realize the *actual signal frequency* has not changed—only the scale factor by which it is displayed. Like the vertical vernier controls, this control allows you to make comparative measurements between signals when you are *not* specifically interested in the *actual signal frequency or period* (for example, phase difference, which is measured in degrees).

    Once again, *it should be clear that you have no way of knowing the actual period of a signal displayed under these conditions, because the vernier is out of its calibrated (CAL) position.*

10. Restore the control to its **CAL** position. The same control may be labeled with the designation × **10.** If so, pull out the knob, and watch the effect on the

display. The control produces a magnification of ten in the sweep speed; that is, the sweep is ten times *faster* when this control is pulled out than when the control is in its default position (pushed in). To get the *actual horizontal attenuator setting,* when this control is activated should you *multiply* or *divide* by ten the setting on the **SECS/DIV** switch?

11. While at the ×10 setting, turn the horizontal **POSITION** control to the left, and to the right, and notice that you are able to see, in *much greater detail,* the voltage variations along the waveform itself. Restore the horizontal vernier control to its normal (×1) position.

    The previous steps would apply equally well if you were using **Ch. 2.** If you wish, and for additional practice, you might repeat these steps using the **Ch. 2** controls.

12. Try using these techniques to measure the p–p amplitude and period of some randomly chosen ac voltages, recording this information in Table 30-1. Columns are provided for recording the vertical sensitivity V (**VOLTS/DIV**) and horizontal sensitivity H (**SECS/DIV**), as well as the amplitudes (V_{p-p}) and periods (T) that you calculate. You might also sketch the form of each waveform alongside the respective row in which you record the data.

DATA FOR EXPERIMENT 30

TABLE 30-1

V (**VOLTS/DIV**)	Def (**DIV**)	Amp. (V_{p-p})	H (**SECS/DIV**)	Def (**DIV**) T (**SEC**)

NOTES

QUESTIONS FOR EXPERIMENT 30

1. A waveform is known to have a peak-to-peak value of 6 V. If it is viewed on the screen with the vertical sensitivity setting at 2 **VOLTS/DIV,** how many major divisions (in the vertical direction) would it occupy?

( )   **(a)** 2   **(b)** 1   **(c)** 3   **(d)** 4

2. Continuing with question 1, if the vertical sensitivity were changed to 1 **VOLT/DIV,** how many divisions would the waveform now occupy?

( )   **(a)** 2   **(b)** 3   **(c)** 4   **(d)** 6

3. If one period of this waveform occupied a distance of five major divisions across the screen, and the horizontal sensitivity were set to 2 **ms/DIV,** what is the frequency of the waveform?

( )   **(a)** 10 Hz   **(b)** 10 kHz   **(c)** 100 Hz   **(d)** 1 kHz

4. Suppose that the horizontal sensitivity control is set to 10 **ms/DIV** and the ×10 multiplier associated with this control is activated, then what is the effective **SECS/DIV?**

( )   **(a)** 1 **ms/DIV**   **(b)** 100 **ms/DIV**   **(c)** 10 **ms/DIV**

5. Why must you have the vertical and horizontal verniers in their **CAL**ibrated positions when making actual voltage and time measurements?

6. Assuming that one side of the function generator is *earthgrounded,* what would happen if you inadvertently connected the *grounded* side of the oscilloscope probe to the *nongrounded* side of the function generator? You might draw a sketch to aid your answer.

31

FUNCTION GENERATOR FAMILIARITY, I

REFERENCE READING

Principles of Electric Circuits: Sections 11–8 through 11–10.

RELATED PROBLEMS FROM *PRINCIPLES OF ELECTRIC CIRCUITS*

None.

OBJECTIVES

To become familiar with the basic operating controls of the function generator. To learn how to set up a waveform of given peak-to-peak value. To learn how to use the dc offset control.

EQUIPMENT

Dual-trace oscilloscope and 1× probes
Audio function generator with square-, triangle-, and sine-wave outputs
DMM

BACKGROUND

The signal, or function, generator is a source of ac waveforms commonly used in electronics. Typically, instruments may generate a variety of signals, including the sine wave, square wave, and triangular wave. All have a level (amplitude) and frequency control as a minimum. These controls are not generally calibrated, and so instruments such as the DMM, oscilloscope, and frequency counter must be used to accurately set the level and frequency. More elaborate generators have a number of extras, such as rectangular (pulse) outputs with a variable duty cycle, which adjusts the relative amounts of time the wave spends at each of two voltage levels. Some generators have push-button attenuators, which can be used to cut the waveform amplitude by a fixed, calibrated amount, such as a factor of ten, twenty, or one hundred. If your generator has attenuators, we will take a look at their effects in this experiment.

In addition to an amplitude control, many generators have what is called a **DC OFFSET** control on them. This allows the user to "shift" the waveform in the positive or negative direction, thereby adding what is called a *dc component* to it. You will see in a later experiment that the dc component is the same as the average value of the waveform, but, for now, we are going to limit our experience to watching the effect of this control on the displayed signal on the oscilloscope screen. In this experiment, we are also going to become more familiar with the **AC/DC** (input coupling) switch on our oscilloscope.

If the particular generator that you are using does not have a dc offset control, then simply ignore those parts of the experiment that refer to it.

PROCEDURE

1. Examine the front panel of your function generator, and locate and identify the following controls. (Note that these are typically on many generators, but, depending on the particular generator that you are using, not all of these controls may exist.)

   (a) **AMPLITUDE** control—adjusts the *peak-to-peak* value of the output signal voltage.

   (b) Fine **FREQUENCY** control—provides continuous control of the frequency.

   (c) **DECADE FREQUENCY** switches—provide frequency selection range by decades (factors of ten).

   (d) **DC OFFSET** control—provides for the adjustment of the dc component (average value) of the output signal voltage.

   (e) **WAVEFORM SELECTOR** switch or switches—allow(s) the user to select the waveform at the output (sine, square, triangle, etc.) of the generator.

   (f) **SYMMETRY** or **DUTY CYCLE**—provides for the adjustment of the pulse width for a pulse (rectangular) output. Some function generators do not have this capability.

   (g) **ATTENUATORS**—provide for the attenuation of the output signal voltage in fixed amounts.

Function Generator Connections and Grounding

2. Use the DMM on **OHMS** to find out if either of the generator's output terminals is earth-grounded. Also check to see if the chassis of your instrument is connected to earth ground. It often is for safety purposes. If it turns out that one of the output terminals of the generator *is earth-grounded* through the line cord, then it is important to keep in mind that this terminal is *always* earth-grounded; this means that your *oscilloscope earth ground must always be connected to this point. Even if you do not make this connection, the two points will be connected through the supply lines.*

3. Set up the oscilloscope to view a signal on channel 1. Set up a ground reference trace on channel 1. Hook up this channel to the generator output terminals, and switch on the generator.

4. Connect the generator output terminals to the oscilloscope. (If one of the function generator's output terminals is marked **LOW** or has a ground symbol, *or* if you found that one of the terminals was actually earth-grounded, then hook it up so that *this terminal* of the function generator is connected to the ground clip of the oscilloscope probe.)

   Note: If your function generator *has* a dc offset control, proceed to step 5; otherwise go directly to step 6.

5. Disable (switch off) any **DC OFFSET** control the generator may have. The *dc component* (or level) of the output signal should now be 0 V. This means that there should be no *dc* in the waveform that is being produced by the function generator. You can confirm this by measuring the voltage across the output terminals, using the DMM set to **DC VOLTS.** The meter might indicate a small "residual" offset voltage and if so, you should use the **DC OFFSET** control to tune it out. This is sometimes called "nulling the offset."

6. Set the generator for a sine-wave output of any amplitude at about 1 kHz. If the generator has any switchable attenuators, they should be *disabled* (switched off) at this point.

7. With the oscilloscope input coupling switch on **DC,** confirm that the waveform is symmetrical about 0 V. (Recall that you nulled the dc offset in step 5.)

8. Switch, in turn, to the other wave shapes that the generator is capable of generating. In each case, the output amplitude and frequency should remain the same; only the kind of signal should change.

   If your function generator has no dc offset control, go directly to step 12.

dc Offset, Range, and Effect on Output Voltage

9. Restore the output to a sine wave. With the oscilloscope connected, place the DMM on **DC VOLTS** across the output. Enable (switch on) the **DC OFFSET** control, and turn it in the direction that increases the offset *positively.* As you do so, the waveform displayed on the oscilloscope should rise so that it is no longer symmetrical with respect to ground. At the same time, the reading on the DMM should increase through *positive* values. Turn it in the opposite direction, and notice that the dc level of the output voltage waveform goes down and the DMM reading becomes *negative* and increases in *magnitude.* Restore the offset to 0 V.

"Clipping" Effect on Some Function Generators

10. Increase the amplitude of the sine wave to its maximum value. If you now increase the offset voltage you *might* notice that the top portion becomes "clipped." Likewise, as you decrease the offset, it might clip at the most negative values of the output signal. If it happens at all, it is more likely when the dc offset and amplitude controls are *both* set for very large outputs. Not all signal generators do this, so do not be alarmed if you cannot get your particular generator to "clip." Just go on to the next step.

    The reason this clipping occurs has to do with internal saturation of the amplifiers that are used to drive the output. Under normal circumstances, this represents unwanted output distortion and is to be avoided. How will you avoid this condition? Would you know that it was happening if you did not have the oscilloscope connected; i.e., would a dc meter necessarily tell you that this was occurring?

11. Use the oscilloscope to determine the peak voltage at which the output clips positively. Repeat for negative clipping. They are normally about equal and opposite in polarity.

Setting Up a Signal of Given p–p Value

12. How do you set up a signal with a given peak-to-peak amplitude? Let's say you want a 1 kHz triangle wave of 2 V p–p. This describes a voltage waveform that oscillates between minimum and maximum values of -1 V and $+1$ V, respectively.

13. Switch off the **DC OFFSET** control or null it to 0 V as shown earlier. Set the **SECS/DIV** to 1 **ms/DIV** and the **VOLT/DIV** to 1 **VOLT/DIV.**

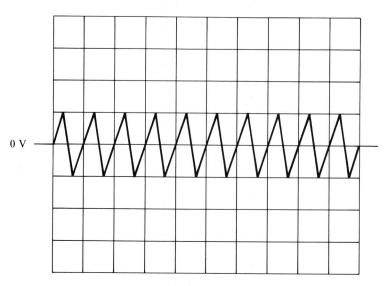

FIGURE 31-1

14. Use the oscilloscope to, as best as you can, set up a triangle wave of 2 V p–p. It should be symmetrical about the ground reference trace (going between the values of −1 V and +1 V). The amplitude control is, of course, used to set up the p–p value.

15. Adjust the frequency dial on the generator until you have exactly ten cycles of the waveform displayed across the screen, as shown in Figure 31-1.

16. With ten cycles of the waveform in ten major divisions, each division represents one period of the waveform. Each period is therefore 1 ms, which means that the frequency is 1 kHz.

  If your generator does not have a dc offset control, go directly to step 21.

Setting Up a Signal of Given p–p Value and Offset Value

17. Suppose that the waveform that needs to be set up must have the same peak-to-peak value, but in addition, have a *dc offset* of 1 V. Enable the **DC OFFSET** control, and adjust it until the ac signal you have displayed is *raised up* by 1 V. The triangle wave should be sitting on top of the 0 V line and should have a maximum value of +2 V, as shown in Figure 31-2. You can then use the DMM (set to **DC VOLTS**) to set the offset more precisely. Extended use of the dc offset control will be studied in a later experiment on pulse and rectangular waves.

18. If you switch from **DC** to **AC** coupling on the oscilloscope channel you are using, you will be able to see the effect of the coupling switch on the displayed waveform. Even though the signal has a dc offset of 1 V, the oscilloscope is displaying the signal as though it didn't. *This is because, on **AC** coupling, only the ac component (the changing part of the signal) is displayed on the oscilloscope screen.*

19. As a quick means of checking the ac and dc components of a waveform, you can flip quickly between **AC** and **DC** coupling on the oscilloscope; you should see the signal *shift vertically* by an amount of voltage equal to the dc offset value.

20. Restore the output to a symmetrical triangle or sine wave at 1 V p–p.

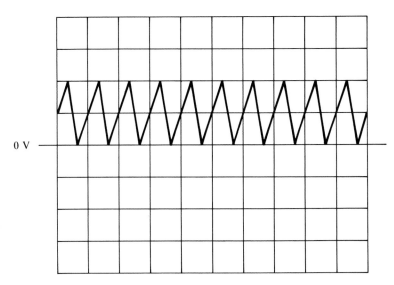

FIGURE 31-2

Use of Attenuators on Some Function Generators (Optional)

21. Locate the attenuator switches on the front panel of your function generator. Your lab instructor will explain the labeling that may appear on these, which may be in decibels (dB) or absolute units.

22. Switch in the attenuators and note their effect on the output waveform. When attenuators are operative, they reduce the amplitude of the waveform by a fixed amount. The amplitude of the displayed waveform should *decrease*. If the waveform disappears from the oscilloscope display, switch to a more sensitive **VOLTS/DIV** setting, and be sure that the triggering is set to **AUTO.**

23. You should experiment with the attenuator controls until you feel comfortable with them. They can be used in later experiments when the output voltage needs to be very small.

QUESTIONS FOR EXPERIMENT 31

   1. If one of a function generator's terminals is earth-grounded, then the oscilloscope ground clip lead
      (a) must be connected to this terminal
      (b) must be connected to the other terminal

( )   **(c)** can be connected to either terminal

   2. The dc offset control on a function generator
      (a) changes the p–p value of the waveform
      (b) changes the dc level or component of the waveform
      (c) does both **a** and **b**

( )   **(d)** does neither **a** nor **b**

   3. Attenuators are used to change
      (a) the amplitude of the output
      (b) the frequency of the output
      (c) both **a** and **b** simultaneously

( )   **(d)** neither **a** nor **b**

   4. If the vertical position of a signal on an oscilloscope screen shifts when you switch from ac to dc coupling, then
      (a) the signal has no dc component
      (b) the signal has no ac component
      (c) the signal has a dc component

( )   **(d)** there is something wrong with the output signal

   5. In your own words, briefly describe the function of two of the following on *your* particular generator:
      (a) the amplitude control
      (b) the smooth (fine) frequency control
      (c) the dc offset control
      (d) any switches that control frequency

   6. In order to see that a waveform has a dc component in it, what form of coupling is necessary to select on the oscilloscope and why?

32

FUNCTION GENERATOR FAMILIARITY, II

REFERENCE READING

Principles of Electric Circuits: Sections 11–8 through 11–10.

RELATED PROBLEMS FROM *PRINCIPLES OF ELECTRIC CIRCUITS*

None.

OBJECTIVE

To become familiar with the basic operating controls of the function generator. To determine the equivalent (Thevenin) output resistance of a generator.

EQUIPMENT

Dual-trace oscilloscope and 1× probes
Audio function generator with square, triangle, and sine-wave outputs
Potentiometer 1 kΩ, or decade box
DMM
Resistors (±5%): 100 Ω (use for a 50 Ω output)
                1 kΩ (use for a 600 Ω output)

BACKGROUND

All function generators have an effective internal resistance, normally referred to as the *output resistance* (sometimes people will call it the output impedance). It is usually quoted as a resistance value and given the symbol R_s or R_o; for example, 600 Ω or 50 Ω is common for audio function generators. *The effect of this resistance is to produce output voltage variations under differing load conditions.* Beginning students often fail to realize this important point. A function generator is *not* a regulated source, and its terminal voltage will vary depending on the load that is connected to it. In this experiment, we are going to improve our familiarity with the function generator; we are going to find both the maximum and minimum (p–p) voltage it is capable of generating when unloaded. Then, using a known load resistor, we are going to try to determine the generator's output resistance. Finally, the "potentiometer method" of finding the output resistance will be used, since this is a common and rather fast way of finding its value. When you have calculated the value of R_s, you should compare it with the value given in the operations manual for the generator. One final point: be sure to find out ahead of time which of the resistors, 1 kΩ or 100 Ω, your instructor wants you to use for this experiment.

PROCEDURE

1. Set up the function generator output for a symmetrical (0 V dc component) triangular wave of *any* amplitude and of frequency 1 kHz. Use the oscilloscope to verify and fine-tune the frequency. Do this with no load connected at this point; that is, with the output terminals open-circuited.

2. By increasing the output amplitude, determine the maximum p–p voltage available across the output terminals under no-load (open-circuited) conditions. Record this in Table 32-1. (*Note:* If your generator exhibits any clipping distortion, as described in experiment 31, record the maximum output voltage *before* the onset of clipping.)

3. Measure and record the *minimum* p–p output voltage across the output terminals. This voltage may be so small that you have to turn the **VOLTS/DIV** on the oscilloscope to the *most* sensitive setting. If the generator has attenuators, you should activate (switch on) these before you make this measurement.

Determination of Generator Output Resistance R_s

Note: Accuracy in measuring R_s of this part of the experiment depends largely on how accurate your two terminal-voltage measurements are, so use as much of the oscilloscope graticule as you are able when measuring the voltages at the function generator terminals.

4. Use the oscilloscope to set the open-circuit voltage to a value of 2 V p–p at the 1 kHz frequency. (The oscilloscope is a very high-resistance device and therefore can be treated as an open circuit to the function generator.)

5. Consult with your instructor and connect *either* the 1 kΩ or 100 Ω load (R_L) across the output terminals, as shown in Figure 32-1. What do you see has happened to the peak-to-peak voltage? You should have seen the voltage decrease from its open-circuit value. The amount of decrease will depend upon the internal resistance of your particular function generator. Measure and record in Table 32-1 this new value of terminal voltage.

6. Calculate and record in the table the percentage drop in output voltage when the generator was loaded. The percentage change can be calculated by

$$\Delta V\% = (V_{O/C} - V_{loaded})/V_{O/C} \times 100\%$$

where $V_{O/C}$ is the open-circuit voltage and V_{loaded} is the voltage under load.

7. Using Ohm's Law, calculate and record in the table the current I_L that flows through the load R_L under these conditions. Then use Kirchhoff's Voltage Law to calculate the voltage drop V_{R_s} across the *internal resistance R_s under loaded conditions.*

8. Now, using the known voltage drop across R_s and the current through it, use Ohm's Law to calculate and record the value of R_s. This is the effective "inter-

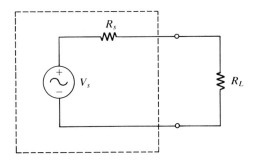

FIGURE 32-1

nal" or "output" resistance of the function generator. Compare the calculated value of R_s with the specified value. (The output resistance is often silk-screened on the front panel of the generator, or you can get it from the operations manual.)

Getting R_s Using the Potentiometer Method

9. Leaving the amplitude control untouched, remove the resistor and replace it with the 1 kΩ potentiometer, so that you are able to vary the load as the wiper is moved. Notice that the terminal voltage is, again, less than its open-circuit value, which was 2 V p–p.
10. Adjust the potentiometer, and observe the terminal voltage. It should vary as the wiper on the potentiometer is moved up and down. Can you see why this should be the case?
11. Carefully adjust the potentiometer so that the terminal voltage *falls by a factor of exactly two* compared with its open-circuited value. When this happens, what do you know about the resistance of the potentiometer in relation to that (internal) of the generator?
12. Carefully remove the potentiometer and, using your DMM on **OHMS,** measure its set value. It should be close to the specified output resistance of the function generator and should compare well with the value determined in step 8.
13. If you refer to the operations manual, you should see both open-circuit (un-loaded) terminal voltages and voltages into a resistance equal in value to the internal resistance R_s. You should now understand the significance of these terms. When the generator is connected to a load equal in value to its output resistance, the voltage falls by a factor of two.

DATA FOR EXPERIMENT 32

TABLE 32-1

Maximum output (volts p–p)	
Minimum output (volts p–p)	
$V_{O/C}$ (volts p–p)	2 V
V_{loaded} (volts p–p)	
ΔV (%)	
I_L (mA p–p)	
V_{R_s} (volts p–p)	
Calculated R_s (step 8)(Ω)	
Measured value R_s (step 12) (Ω)	

NOTES

QUESTIONS FOR EXPERIMENT 32

1. A source similar to the one used in this experiment has a value R_s of 50 Ω and a maximum open-circuit terminal voltage of 7 V peak. The value of a load that will cause it to fall to 5 V is closest to

( )   **(a)** 50 Ω   **(b)** 125 Ω   **(c)** 20 Ω   **(d)** 70 Ω

2. A source having an output impedance of 600 Ω is to be connected to a load so that the terminal voltage drops by no more than 5 percent of its open-circuit value. The minimum value of the load is

( )   **(a)** 11.4 kΩ   **(b)** 32 Ω   **(c)** 570 Ω   **(d)** 57 kΩ

3. A source similar to that used in this experiment and having a value R_s of 600 Ω is found to have a minimum open-circuit terminal voltage of 500 mV. The user requires a voltage of 50 mV. This can be accomplished by placing a resistor across the output terminals. Its value is

( )   **(a)** 60 Ω   **(b)** 67 Ω   **(c)** 5.4 kΩ   **(d)** 54 Ω

4. A source having an output impedance of 600 Ω is required to be reduced to 100 Ω. The value of a terminal resistor that will accomplish this is

( )   **(a)** 100 Ω   **(b)** 120 Ω   **(c)** 500 Ω   **(d)** 85 Ω

5. An experiment with changing load conditions requires a constant voltage across this load. What can be done to ensure that the load voltage remains constant as the load changes?

6. Many students attempt to "measure" R_o by positioning an ohmmeter directly across the generator terminals. Why is this inadvisable no matter whether the generator is switched on or off?

182

33

AMPLITUDE VALUES OF A SINE WAVE

REFERENCE READING

Principles of Electric Circuits: Sections 11–2 through 11–3.

RELATED PROBLEMS FROM *PRINCIPLES OF ELECTRIC CIRCUITS*

Chapter 11, Problems 8, 9, and 10.

OBJECTIVE

To be able to distinguish among rms, peak, and peak-to-peak values of a sine wave.

EQUIPMENT

Audio signal generator
Oscilloscope and 1× probe
DMM or VOM

BACKGROUND

There are a number of ways to describe a periodic waveform. The three most common are the peak value, the rms value, and the average value. The peak values tend to be used more often in analog electronics and in conjunction with amplifiers. The rms value is more often used in power systems, particularly where information about energy usage is important. The average value is perhaps not as commonly used as the other two, although it is fundamental in the areas of filtering and rectification. What follows is a summary of the definition of the three values. The peak value is the maximum excursion of the signal measured from a zero reference. The rms value is the value of a dc voltage (or current) that would have the same heating effect (equal power) on the average as that of the signal. Finally, the average value is best thought of as that which a dc meter would read if placed in the circuit. Notice that the dc meter does *not* measure the rms value.

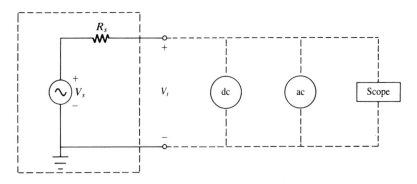

FIGURE 33-1

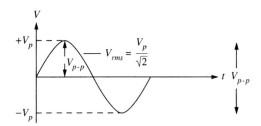

FIGURE 33-2

Of particular interest with a sine wave is the half-cycle average value, which is the reading a dc meter would give if it were inserted into a circuit when the negative lobe of the sine wave had been inverted. Its usefulness will become evident later. Finally, the peak-to-peak value of a sinusoid is twice its peak value.

PROCEDURE

1. Connect the meter to the generator terminals. (Be sure it is set for ac volts.)
2. Adjust the generator output to provide a 1 V rms sine wave at a frequency of about 1 kHz.
3. Disconnect the meter and connect the oscilloscope in its place (Figure 33-1). Measure and record the peak and peak-to-peak values in Table 33-1. Be sure to maximize use of the screen so as to get better accuracy.
4. Now set the meter to read dc.
5. Connect the meter back into the circuit where you have a 1 V rms sine wave. Record the reading indicated by the meter in Table 33-1 under Measured Average Value.
6. Repeat steps 1 through 5 for a 2 V rms signal, then a 3 V rms signal.
7. Reconnect the oscilloscope and adjust the generator output to provide 1 V peak. Record the measured peak-to-peak value in the table (Figure 33-2).
8. Return the meter to the circuit and record the rms value.
9. Calculate and record the theoretical *half-cycle* average value. (You cannot measure this at this point.)
10. Repeat steps 7 through 9 for 2 V and 3 V peak signals.
11. Set the generator output to any convenient value and observe the signal with the oscilloscope.
12. Slide the oscilloscope coupling switch from **DC** to **AC** and back again. If the waveform has no dc component, then the waveform should not shift upward or downward.

Name _____ Date _____

DATA FOR EXPERIMENT 33

TABLE 33-1 *Amplitude values*

rms	Measured Peak	Measured Peak-to-peak	Measured Average Value	Calculated Half-cycle Average
1 V				
2 V				
3 V				
	1 V			
	2 V			
	3 V			

NOTES

QUESTIONS FOR EXPERIMENT 33

( )
1. A sine wave has an rms value of 1.4 V. Its peak value is
   (a) 1.4 V   (b) 2.8 V   (c) 2.0 V   (d) 1 V

( )
2. The half-cycle average value of the signal in question 1 is
   (a) 1.3 V   (b) 0.89 V   (c) 0.64 V   (d) 1.8 V

( )
3. A sine wave has a peak-to-peak value of 1 V. Its rms value is
   (a) 0.35 V   (b) 0.71 V   (c) 1.41 V   (d) 0.32 V

( )
4. A sine wave has a *half-cycle* average value of 1.0 V. Its rms value is
   (a) 1.5 V   (b) 2.2 V   (c) 1.1 V   (d) 0.55 V

5. Explain in your own words why the ac/dc switch on your oscilloscope had no observable effect on the position of the signal on the screen in step 12.

6. Discuss the advantages and disadvantages of using the VOM versus the oscilloscope to determine both the rms and peak values.

34

PHASE MEASUREMENT

REFERENCE READING

Principles of Electric Circuits: Section 11–4 and Chapter 12 TECH TIP.

RELATED PROBLEMS FROM *PRINCIPLES OF ELECTRIC CIRCUITS*

Chapter 11, Problems 11 through 20.

OBJECTIVE

To measure the phase difference between two sinusoidal voltages using a dual-channel oscilloscope.

EQUIPMENT

Audio signal generator
Oscilloscope and two 1× probes
Resistor (±5%): 16 kΩ
Capacitor: 0.001 μF

BACKGROUND

When two sine waves of the same frequency are compared, their peak values and zero crossings, will, in general, not occur together. The signals are said to be *out of phase*. The amount by which one signal is ahead or behind the other is measured in either units of time or units of angle. The angular units may be radians or degrees, though the latter is generally preferred.

The oscilloscope can conveniently measure the phase difference between two signals of the same frequency. Using the dual-trace feature, both signals are displayed with the same ground reference. The horizontal vernier is then taken out of **CAL**ibrate so that a cycle (or half-cycle) of the signal can be fitted into a whole number of divisions across the screen. The divisions can then be scaled off in terms

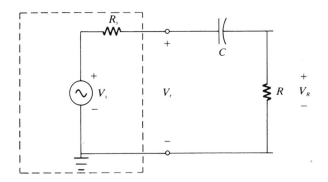

FIGURE 34-1

of angle and the phase difference computed. The distance between consecutive corresponding zero crossings of the signal is normally used for this purpose. With a knowledge of the frequency, the time difference corresponding to the angular difference can be computed.

In this experiment, you will become familiar with two more oscilloscope controls, **LEVEL** and **SLOPE.** These controls can be used to set up where on the observed waveform the sweep begins. You should experiment with these controls and try to understand just what it is they can do for you.

PROCEDURE

1. Connect up the circuit shown in Figure 34-1.
2. Set up the oscilloscope for dual-channel operation (for viewing two signals simultaneously).
3. Use the input coupling switches and the vertical **POSITION** controls to set up ground reference traces for both channels. Position channel 1 two divisions (cm) from the top and channel 2 two divisions from the bottom of the graticule.
4. Set both vertical timebases at 0.5 **V/DIV,** and set up the oscilloscope triggering to trigger from channel 1. Set the horizontal attenuator to 20 µs**/DIV.**
5. Connect channel 1 of the oscilloscope across the terminals of the generator and adjust for a sine-wave output of 2 V p–p at 10 kHz (zero dc component).
6. Connect channel 2 of the oscilloscope across the resistor. At this frequency, the resistor voltage will have an amplitude somewhat smaller than that at the input to the circuit and will be out of phase with it.
7. Set the input coupling switches to **GROUND,** and use the vertical **POSITION** controls to bring the traces to the horizontal center line of the graticule.
8. Restore both coupling switches to **DC,** so that you are again able to see both V_t and V_R on the screen.
9. Temporarily make adjustments so that only channel 1 is visible on the screen. Use the horizontal **POSITION** control to bring the far left portion (beginning) of the waveform into view.
10. Set the trigger mode switch from **AUTO** to **NORMAL.** This will allow you to control the triggering point of the signal on channel 1 with the **LEVEL** and **SLOPE** controls.
11. Now use the trigger **LEVEL** and **SLOPE** controls to set up the display so that the sweep begins when the sine wave is going through 0 V and increasing (see far left of Figure 34-2).
12. By switching to 10µs**/DIV,** and taking the horizontal vernier out of its **CALi**brated position, you should be able to spread out the waveform so that one full

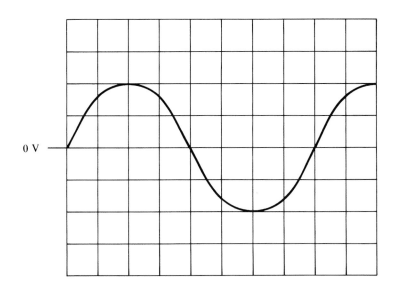

0 V

FIGURE 34-2

cycle takes up eight divisions across the screen. When you have completed this step, the display should look like that shown in Figure 34-2. How many degrees does a major division (1 cm) now represent? How many degrees does a minor division (0.2 cm) represent?

13. Bring channel 2 back into view. Notice that at the instant V_t (on channel 1) passes through 0 V, at the far left of the screen, V_R (on channel 2) is actually at some positive voltage.

14. You should be able to see whether the voltage on channel 1 leads or lags that on channel 2. Record this in Table 34-1.

15. Determine by how many degrees these two signals are out of phase. Measure the distance in major divisions between *corresponding zero crossings* of the two waveforms, and then, using the relationship you developed in step 12, convert this to degrees.

16. Because 360° represents one period of this waveform, and 10 kHz has a period of $1/10^4 = 100$ μs, you can, by using ratio and proportion, calculate how much time in microseconds the phase difference corresponds to. Calculate and record this in Table 34-1.

17. Maintaining the amplitude of the input at 2 V p–p, repeat all of the steps at the remaining frequencies in the table. Note that because the frequency is being changed, it will be necessary to alter the **SECS/DIV** setting to make these phase and time measurements. You should select whatever horizontal timebase settings make your measurements the most convenient.

Name _____ Date _____

DATA FOR EXPERIMENT 34

TABLE 34-1

Frequency (kHz)	Phase (°)	Δt (µs)	V_R lags or leads V_t
10			
5			
50			

NOTES

QUESTIONS FOR EXPERIMENT 34

1. In a circuit similar to that in Figure 34-1, a time difference of 10 μs at 10 kHz represents a phase shift of about
( )   (a) 20π rad   (b) 3600°   (c) 36°   (d) 3.6 rad

2. At what frequency would a time delay of 0.5 μs be equivalent to 90° of phase shift?
( )   (a) 2 MHz   (b) 1 MHz   (c) 500 kHz   (d) 250 kHz

3. A given amount of phase shift represents a greater time delay at
( )   (a) higher frequencies   (b) lower frequencies

4. In this circuit, the resistor voltage was found to
        (a) always lead the input voltage
        (b) always lag the input voltage
( )   (c) sometimes lag and sometimes lead the input voltage

5. Describe in your own words why it is convenient to use an uncalibrated time base to measure phase difference. Why is the actual time itself of no concern?

6. Using ratio and proportion, can you write a formula that will give the time difference Δt between the two sine waves, when you know the phase difference Φ and the frequency f? Write a formula involving Δt, Φ, and f, with the subject of the formula being Δt. (*Hint:* How did you determine the time difference in step 16 of the experiment?)

35

NONSINUSOIDAL WAVEFORMS

REFERENCE READING

Principles of Electric Circuits: Section 11–8.

RELATED PROBLEMS FROM *PRINCIPLES OF ELECTRIC CIRCUITS*

Chapter 11, Problems 29 through 38.

OBJECTIVE

To examine various kinds of pulse waveforms and observe their properties.

EQUIPMENT

Pulse waveform generator (with square-wave output)
Oscilloscope and $10\times$ probe
dc power supply (if no dc offset control on waveform generator)
DMM or VOM
Capacitor: 0.001 µF

BACKGROUND

In this experiment you will become familiar with nonsinusoidal waveforms—in particular, square and pulse types. These kinds of signals are predominant in digital and computer systems.

The first part of the experiment deals with the rise/fall time of waveforms. Though books might draw square waves with abrupt vertical transitions, in reality all transitions take some time to occur. The standard definition of rise time is the time it takes the waveform to go from 10 to 90 percent of its total (p–p) amplitude. This measurement is so common that many oscilloscope manufacturers have silk-screened these markings on the CRT graticule. Because the rise time of the generator might be too fast for you to measure easily, in this experiment a capacitor connected across the output terminals will serve to artificially slow down the transitions.

The remainder of the experiment is concerned with the determination of standard characteristics of pulse waveforms. The definitions of these follow so that you can refer to them during the procedure.

Baseline—the most negative voltage in the waveform
Amplitude—in this case, the peak-to-peak amplitude
Period—the time for one cycle of the waveform
PRF—the pulse repetition frequency, the number of cycles per second
Pulse width—the time over which the most positive voltage exists
Duty cycle—the ratio of the pulse width to the period
Average value—the value a dc meter will indicate

PROCEDURE

Part A: Measurement of Rise Time of a Square Wave

1. With its output connected to the channel 1 input of the oscilloscope, set up the function generator for a symmetrical (no dc component) square-wave output at some convenient amplitude, say 2 V p–p and frequency 10 kHz. Be sure that you set up a ground reference on the oscilloscope screen, and switch back to **DC** coupling for viewing the waveform.
2. Connect the capacitor directly across the output terminals of the generator. This serves to artificially slow down the rise time so that you can easily measure it without using any of the advanced delay features of the oscilloscope.
3. Take the channel 1 vernier out of **CAL,** and use it to fit the peak-to-peak amplitude of the signal into exactly five major divisions spaced equally around the horizontal center (ground reference) line. This means that the most positive level of the waveform will extend 2.5 divisions above the center and the most negative (baseline), 2.5 divisions below the center. Change the **VOLTS/DIV** control if necessary.
4. Set the **SECS/DIV** control to 0.1 μ**s/DIV.** Use the horizontal **POSITION** and trigger **LEVEL** and **SLOPE** controls to bring a *single rising edge of the square wave into view* at the far left-hand side of the screen. The oscilloscope will need to be set to **NORMAL** triggering to do this.
5. The rise time of the waveform is the time taken for the rising edge to go from 10 to 90 percent of its p–p amplitude (Figure 35-1). Because the signal is spread over 5 divisions, *10 percent of the p–p amplitude is one-half of a major division, and 90 percent is four and one-half major divisions.* Use this fact to measure the rise time of the rising edge of the square wave, and record your data in Table 35-1. Some oscilloscopes have 10 and 90 percent markings silkscreened onto the graticule in these positions.
6. Actuate the trigger **SLOPE** control so that the sweep begins on the falling edge of the square wave. You will have to rotate the **LEVEL** control to bring the falling edge of the waveform into view at the far left of the screen.
7. Now repeat the above procedure and measure the fall time of the square wave. Record the data in Table 35-1. The rise and fall times may differ somewhat, but should be about the same order of magnitude.
8. Remove the capacitor from the terminals of the function generator before you continue with the experiment.

Part B: Square-Wave Measurements

1. With its output connected to the channel 1 input of the oscilloscope, set up the function generator for a symmetrical (no dc component) square-wave output

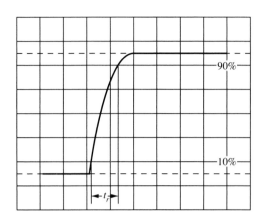

FIGURE 35-1

at some convenient amplitude, say 4 V p–p and frequency 1 kHz. Be sure that you have set up a ground reference on the oscilloscope screen.

2. The oscilloscope should display a symmetrical signal, going equally positive and negative with respect to the centerline.

3. Measure and record the baseline, *peak-to-peak* amplitude, and period in Table 35-2. Then, using the known period, calculate the PRF (pulse repetition frequency) and confirm that it matches the setting on the function generator dial.

4. Carefully measure and record the pulse width t_w.

5. From t_w and T, calculate and record the duty cycle and then the average value of the square wave. The average value of the waveform should, of course, be close to 0 V.

6. Connect the DMM set to **dc VOLTS** across the function generator's output terminals (in parallel with the oscilloscope), and record the indicated value. It should closely match your calculated (average) value from step 5. Switch the oscilloscope input coupling for channel 1 from **DC** to **AC,** and watch the trace. Its position should stay essentially the same—there should be no upward or downward movement. Why? Restore the coupling switch to **DC.**

7. If the dc meter did not read exactly 0 V in step 6, you can often make a slight adjustment of the dc offset control (if one is provided) to bring this reading to 0. If the generator has a dc offset control, continue with step 8. If not, go directly to step 9.

8. Turn the dc offset so that the displayed waveform rises by 2 V. The baseline of the signal should now be at 0 V and the peak value at 4 V instead of 2 V. (Your oscilloscope channel 1 must, of course, be set to **DC** coupling for this operation.) Go to step 10. *Do not alter the p–p amplitude of the waveform.*

9. Take a low-voltage (floating) dc power supply, and connect it in series with the signal generator (Figure 35-2). Be sure to connect this power supply to the *nongrounded side* of the signal generator. Switch on the supply, and adjust it until the displayed signal rises by 2 V. The baseline of the signal should now be at 0 V and the peak value at 4 V instead of 2 V. (Your oscilloscope channel 1 must, of course, be set to **DC** coupling for this operation.) *Do not alter the p–p amplitude of the waveform.*

10. Repeat all measurements in steps 3 through 6, recording data in the table. The only differences between the entries should be those of the baseline and the average values. Switch the oscilloscope input coupling for channel 1 from **DC** to **AC,** and watch the trace. Its position should change—there should be a downward movement of the display. Notice the amount by which the trace moves downward. Can you relate the vertical shift to the average value? Restore coupling to **DC.**

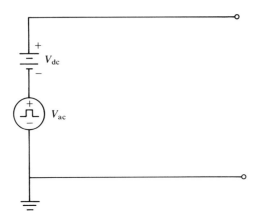

FIGURE 35-2

11. Now insert a 2 V *negative* dc offset so that the square wave is negative going from − 4 V to 0 V. If you are using the power supply method from step 9, simply reverse the polarity of the supply. The power supply must "float" with respect to ground if you are to do this.

12. Repeat all measurements in steps 3 through 6, recording data in the table. The only differences between the entries should be those of the baseline and the average values.

13. Switch the oscilloscope input coupling for channel 1 from **DC** to **AC,** and watch the trace. Its position should change—on this occasion there should be an *upward* movement of the display. Notice the amount by which the trace moves upward. Can you again relate the vertical shift to the average value?

Part C: Pulse-Wave Measurements

Some function generators do not have a pulse (nonsquare) output. If the particular function generator has no pulse output, then omit this part of the experiment.

1. If the function generator has a fixed amplitude and duty cycle pulse output, then set up channel 1 to look at this signal. Adjust the frequency to a value of 1 kHz. If the generator allows you to adjust the amplitude and duty cycle of the pulse, then set up one at some convenient p–p amplitude and duty cycle (not 50 percent). Simply make sure that the pulse width is *not* one-half of the period.

2. Measure and record in Table 35-3 the baseline voltage, peak-to-peak amplitude, pulse width, and period of the pulse wave. Make sure the input coupling for channel 1 is on **DC** for these measurements.

3. From this, calculate and record the duty cycle and the average value of the waveform.

4. Connect the DMM on **dc VOLTS** in parallel with the oscilloscope and record the average value in the table for comparison. It should compare well with the calculated value in step 3.

5. Switch the oscilloscope input coupling for channel 1 from **DC** to **AC,** and watch the trace. Its position should change; depending on whether the average value is positive or negative, there should be either a *downward* or an *upward* movement of the display. Notice the amount by which the trace moves upward. Can you relate the vertical shift to the average value?

Name _____   Date _____

DATA FOR EXPERIMENT 35

TABLE 35-1

Rise Time	Fall Time

TABLE 35-2

Baseline	Amplitude (p–p)	Period	PRF	Pulse Width	Duty Cycle	Average Value	
						Calculated	Measured

TABLE 35-3

Baseline	Amplitude (p–p)	Pulse Width	Period	Duty Cycle	Average Value	
					Calculated	Measured

NOTES

QUESTIONS FOR EXPERIMENT 35

1. A square wave has a baseline of -2 V and peak-to-peak amplitude of 12 V. The most positive voltage in this waveform is

( )   (a) 12 V   (b) 14 V   (c) 4 V   (d) 6 V   (e) 10 V

2. Continuing with question 1, this waveform has an average value of

( )   (a) 12 V   (b) 10 V   (c) 6 V   (d) 5 V   (e) 4 V

3. A dc meter reading of a square wave measures 1 V. It has a baseline of -4 V. What is its peak-to-peak amplitude?

( )   (a) 6 V   (b) 5 V   (c) 10 V   (d) 12 V   (e) 8 V

4. How much of a dc offset voltage needs to be added to the waveform in question 3 to make the dc meter read zero? (Pay attention to sign.)

( )   (a) $+1$ V   (b) $+4$ V   (c) -2 V   (d) -4 V   (e) -1 V

5. Sketch two square waves, one with an average value of zero and one with an average value of 2 V. Label *all* pertinent amplitude and time quantities.

6. Sketch two waveforms with the same duty cycle but different frequencies. Be sure to label all time and amplitude values.

36

CAPACITANCE FUNDAMENTALS

REFERENCE READING

Principles of Electric Circuits: Sections 12–1, 12–2, and 12–5.

RELATED PROBLEMS FROM *PRINCIPLES OF ELECTRIC CIRCUITS*

Chapter 12, Problems 14 through 19.

OBJECTIVE

To test capacitor integrity with an ohmmeter. To observe the basic charge/discharge effect of a capacitor.

EQUIPMENT

dc power supply 0–15 V
VOM or other analog meter
SPDT switch
Capacitors: selection of various capacitors from 0.1 μF through 500 μF
Lamp no. 47 or equivalent (incandescent)

BACKGROUND

Capacitors find many applications in electronic circuits. In this experiment, we will see how a simple ohmmeter test can reveal the condition of a capacitor. We will also discover the basic behavior of a capacitor as an energy storage device in a dc circuit.

Because a capacitor is formed of an insulating medium known as a dielectric, direct current cannot flow between the electrodes. Consequently, an ohmmeter test should reveal an open circuit (∞). In large-valued capacitors, it may take a visibly long enough time for the reading to stabilize as the device charges. That is, the ohmmeter may initially read 0 Ω and then swing to ∞ Ω after a short time. With small-

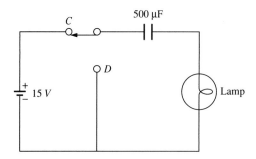

FIGURE 36-1

valued capacitors, however, the effect is almost instantaneous. Some electrolytic capacitors have imperfect dielectrics and will conduct a small amount of current referred to as *leakage*. This shows up as a high resistance on the ohmmeter test. To determine the leakage (if any) of high-quality capacitors, a more sensitive instrument is required.

To demonstrate the "transient" that occurs when a capacitor is connected to a dc source, we will use a lamp. The lamp will light when current is flowing. The switching arrangement shown in Figure 36-1 allows us to charge and discharge the capacitor through the lamp. If the charge/discharge current is large enough and long enough, the lamp should flash momentarily. The experiment is designed to demonstrate this phenomenon. The first part of this experiment works better with analog meters, so, if you can, avoid the use of a DMM here.

PROCEDURE

1. Determine the value of each capacitor from its markings. Record these values in Table 36-1, listing them in order of increasing value. Some capacitors use a code like the resistor color code, using numbers instead of colors directly; the value is usually in pF; e.g., 102 would indicate 10×10^2 pF = 1000 pF. Other capacitors simply have the value stamped on them; e.g., 0.1 M will mean 0.1 μF. (M is often used in place of μ on capacitor bodies.)

2. Use either a capacitance tester or impedance bridge to measure the actual capacitance of the capacitors and record these measured values in Table 34-1. You may need to refer to the operations manual for the tester/bridge, or consult with your instructor for the procedure. In any case, always connect capacitors directly across the terminals of the tester/bridge or else keep the test leads short. Capacitors have wider tolerances than resistors, so expect some variation.

3. Using a VOM on the largest **OHMS** range (always zero the range before using), place the leads across the smallest capacitor, and measure its resistance. You may see a short "kick" on the meter, but the reading should reach a steady value within a short period of time. If the resistance is measurable (not infinity), record the resistance in Table 36-1; otherwise, simply write "open" or ∞ Ω.

4. Repeat step 3 for each of the capacitors in turn. If any of the capacitors are polarized, as are most electrolytic types, be sure that you use the correct ohmmeter polarity—positive side of the ohmmeter to the positive side of the capacitor.

5. Connect up the circuit in Figure 36-1, making sure that, if polarized, the capacitor is in the circuit the right way around. Adjust the source voltage to 15 V. If the lamp is permanently on, you have miswired the circuit; immediately turn off the power supply and check your wiring.

6. Beginning with the switch in the D (discharge) position, throw the switch to position C (charge), and observe the lamp. It should momentarily light up, then remain extinguished.

7. Return the switch to the D position, and the lamp should briefly light again.

8. Flip the switch back and forth from C to D and notice the flickering of the lamp. Try to relate the behavior of the lamp to the charging and discharging of the capacitor.

9. Try flipping the switch back and forth in rapid succession. If you can do this fast enough, the lamp might look like it is permanently on. Note that you are able to produce an almost continuous light even though dc cannot flow through a capacitor. What is the nature of the current that flows in this circuit when the switch is moved in this manner?

10. Beginning once more in the discharge position, use either a VOM or DMM to measure the (steady) voltage across the lamp (V_L) and the capacitor (V_C). Record these values in Table 36-2.

11. Remove the meter from the circuit, and throw the switch back to C. *After the lamp has extinguished,* use the meter to measure the lamp and capacitor voltage, recording the data in the table. Do not keep the meter across the capacitor for more than enough time to take the measurement. A "sample and hold" meter may be used for this measurement. Ask your instructor.

12. With the switch in position C, reconnect the meter across the capacitor and watch the reading. Now disconnect the positive power supply lead from the switch, and watch the meter reading. It should slowly decay toward zero. Why does this happen?

DATA FOR EXPERIMENT 36

TABLE 36-1

Capacitor Details		Measured Resistance	
Coded Value	Measured Value	1st Range	2nd Range

TABLE 36-2

	Switch in Position	
	D	C
V_L		
V_C		

NOTES

QUESTIONS FOR EXPERIMENT 36

( )

1. When tested with an ohmmeter, a good capacitor should yield a resistance close to
(a) $\infty\ \Omega$ (b) $0\ \Omega$ (c) neither of these

2. Which of the following statements is correct for Figure 36-1 when the switch is thrown to position *C?*
(a) Current can flow indefinitely.
(b) Current can flow for a short period of time.

( )   (c) Current cannot ever flow in this circuit.

3. Which of the following statements is correct for Figure 36-1 when the switch is thrown from *C* to *D?*
(a) Current can flow indefinitely.
(b) Current can flow for a short period of time.

( )   (c) Current cannot ever flow in this circuit.

4. With regard to the content of this experiment, the current in the lamp
(a) was always in the same direction
(b) alternated depending on the switch position

( )   (c) was always zero

5. Comment on the amount of time the lamp was on. On what do you think this depends?

6. Referring to step 12 of the procedure, why does the voltage across the capacitor decay when the meter is kept across the capacitor and the power supply is disconnected?

37

RESISTOR-CAPACITOR TIME CONSTANTS

REFERENCE READING

Principles of Electric Circuits: Section 12–5.

RELATED PROBLEMS FROM *PRINCIPLES OF ELECTRIC CIRCUITS*

Chapter 12, Problems 30 through 41.

OBJECTIVE

To become familiar with charge/discharge characteristics of RC circuits. To measure the time constant, and verify the "five time-constant rule."

EQUIPMENT

dc power supply 0–10 V
DMM (high-input impedance)
Capacitors: 0.22 μF, 0.47 μF
Stopwatch or equivalent

BACKGROUND

In the previous experiment, using a simple series circuit containing a lamp, we observed the basic charge/discharge mechanism in capacitors. This experiment examines more closely the forms of the current and voltages in such a circuit. Theory tells that when a dc voltage, such as that in Figure 37-1, is switched across an uncharged capacitor, the voltage across it rises toward the source voltage exponentially. Though in theory the voltage across C never reaches the source voltage, it is practically charged to this voltage within five time constants after the switch has been thrown to the C position. The time constant for the RC series circuit is given by the product of R and C. (Remember C must be in Farads and R in Ohms to get units of seconds.) If we try to verify the charging time using a DMM or some other instrument to measure either the capacitor or resistor voltage, the positioning of the instrument will affect the time constant of the circuit, and we may not get very convincing results.

To overcome the instrument loading effect, in this experiment, we will use the voltmeter itself as the charging resistor. That is, we will place the voltmeter in series with the capacitor, and set it to **dc VOLTS.** This will allow us to very conveniently measure the resistor voltage without any loading effect, since the voltmeter will be measuring the voltage across itself. It will be necessary to know the resistance of the voltmeter so that the expected time constant can be calculated. Knowing the voltage across the meter at any time, indirectly allows us to compute the voltage across the capacitor at the same time since, by Kirchhoff's Voltage Law, they must add up to the source voltage. We will use this idea in the experiment. Since the circuit voltages and currents are exponential in nature we will assume that the capacitor has fully charged if its voltage is within 1 percent of its final value. This means that at the same time, the voltage across the resistor is less than or equal to 1 percent of the source voltage.

One final word. Be sure that the capacitor is fully discharged before taking data on charging curves. In addition, you should take data by averaging the results of several repetitions of the activity.

PROCEDURE

1. From the operations manual or your instructor, determine the resistance R_{in} of the DMM when on **dc VOLTS.** On most DMMs it is 10 MΩ. Record this in each of the rows in Tables 37-1 and 37-2.
2. Use a capacitance tester or impedance bridge to check the capacitance of each of the capacitors, and record these measured values in the table. If neither of these are available, use the nominal capacitance value taken from the body of the capacitor.
3. From the measured values of R_{in} and C, calculate and record the time constant and the expected total charge time (using the five time-constant rule) of the circuit in Figure 37-1.
4. Connect up the circuit in the figure with the switch in the D (discharge) position. Adjust the source voltage for 10 V. The DMM should be set up for **dc VOLTS** on a range that can display 10 V without an overreading. In addition, you should hook up the meter so that the **POSITIVE** side goes to the switch and the **COMMON** side to the capacitor.
5. Throw the switch to the C position. The capacitor will begin to charge through the meter resistance. The voltage across C rises exponentially from 0 V toward 10 V; at the same time, the voltage across R_{in} (indicated by the meter) jumps to 10 V and then decays exponentially toward 0.
6. Once the resistor voltage has reached approximately 0 V, throw the switch to D and watch the meter. The capacitor now discharges back through the meter resistance, so the meter indication should jump to -10 V and then decay exponentially toward 0 V. Why does the meter indicate a negative voltage during the discharge process?
7. Using a stopwatch or other accurate timekeeping device, determine the time it takes for the capacitor to "fully" charge from the moment the switch is thrown to C. Because the capacitor never fully charges, you need to terminate the timekeeping when the voltage across the capacitor has reached approximately 9.8–9.9 V; i.e., when the resistor (meter) voltage has reached about 0.1–0.2 V. *Do not wait for the meter to read 0 V.* You should do this several times and take the average for meaningful results. Be sure to let the capacitor com-

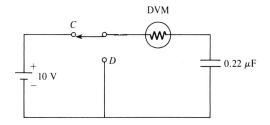

FIGURE 37-1

pletely discharge between trials by throwing the switch back to D and waiting for a complete discharge. Record this as the measured total charge time in Table 37-2.

8. Discharge the capacitor completely by throwing the switch back to D. We will now estimate the time constant of this circuit. Because the voltage across R falls to approximately 37 percent of its initial value during charging, it ought to be possible to estimate the actual time constant (37 percent of 10 V = 3.7 V).

9. Initiate the charging process again, and determine the time it takes the resistor voltage to reach 3.7 V. Once again, do this several times and take the average of the times. Record this average as the measured time constant τ in the table.

10. With the switch initially at C, using the stopwatch, determine the total discharge time when the switch is thrown to D. This will be the time taken for the DMM reading to be close to 0 V; in practice, about 0.1–0.2 V should work. Record in Table 37-2.

11. The discharge time constant can be estimated in the same manner as the charging time constant was. The voltage across the resistor jumps to -10 volts and then decays toward 0 V. Determine the time it takes this voltage to reach -3.7 V from the moment the switch is thrown to D. Take the average of several tries. Record this measured time constant τ in Table 37-2.

12. Repeat the above steps for the remaining capacitors in the tables, recording all charging data in Table 37-1 and all discharging data in Table 37-2.

DATA FOR EXPERIMENT 37

TABLE 37-1

R_{in}	C	τ		Total Charge Time	
		Calculated	Measured	Calculated	Measured

TABLE 37-2

R_{in}	C	Measured Total Discharge Time	Measured τ

NOTES

QUESTIONS FOR EXPERIMENT 37

(   )

1. If the DMM had a value of R_{in} equal to 10 MΩ in this experiment, then with $C = 0.22$ μF, the time constant would have been
   (a) 2.2 ms   (b) 2.2 s   (c) 45.5 ms   (d) 22 ms

(   )

2. With reference to Figure 37-1, using Kirchhoff's Voltage Law, after two charging time constants, the voltage across the meter should be close to
   (a) 3.68 V   (b) 1.35 V   (c) 6.32 V
   (d) 8.65 V   (e) 0 V

(   )

3. With reference to Figure 37-1, using Kirchhoff's Voltage Law, after three discharging time constants, the voltage across the meter and capacitor ought to be
   (a) 4.98 V   (b) 0.498 V   (c) 9.50 V
   (d) 1.35 V   (e) 10 V

(   )

4. When the switch has been at C and the capacitor has therefore charged, no current flows in the circuit. When it is thrown back to D for discharge, the current jumps up to some value and then again decays toward zero. On what does this initial value of current depend?
   (a) R (b) C (c) V_s (d) R and C only
   (e) R and V_s

5. Explain in your own words the reason why the DMM indicated a negative voltage during discharging of the capacitor. With reference to Figure 37-1, which way does the current flow through the meter during discharge?

6. If a resistor had been used instead of the DMM in this experiment, and the DMM had been positioned across the capacitor to observe its charging, what would have been the effect on (a) the time constant and (b) the final voltage to which the capacitor charges?

38

INDUCTORS IN dc CIRCUITS

REFERENCE READING

Principles of Electric Circuits: Sections 13–1, 13–4.

RELATED PROBLEMS FROM *PRINCIPLES OF ELECTRIC CIRCUITS*

Chapter 13, Problems 1 through 5, 19 through 26.

OBJECTIVE

To observe the effects of inductance in dc circuits.

EQUIPMENT

dc power supply 0–10 V
Inductor 0.5 H or above
DMM
Neon lamp
SPST switch

BACKGROUND

In this experiment we will investigate one of the basic properties of inductance, that of being able to generate high voltages by switching off the current supply to the inductor. This process is often used where a high voltage is required but only a low-voltage supply is available. Figure 38-1 illustrates the circuit we will use. The inductor is placed in parallel with the lamp, and this combination is connected to a source by a switch. When the switch is closed, a steady current will be taken by the inductor and the lamp will behave like an open circuit. When the switch is opened, the current is forced through the lamp, making it "fire." A high voltage is required to "fire" the lamp; this voltage is developed by the energy stored in the inductor's magnetic field.

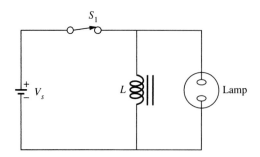

FIGURE 38-1

PROCEDURE

1. Measure the dc resistance and inductance of the inductor using an LCR meter. Record in Table 38-1. Note also its rated current (usually on a name plate, or ask your instructor), if the information is available.
2. Calculate and record the maximum dc voltage permitted across the inductor, using the dc resistance and rated current.
3. Connect the circuit in Figure 38-1. Leave S_1 open for the moment.
4. Adjust the supply voltage to a value of 1 V, then close S_1. Note whether or not the lamp fires. Record this is Table 38-2.
5. Open S_1. Note whether or not the lamp fires on this occasion. Once again, record the lamp's behavior in Table 38-2.
6. Increase the voltage by 0.5 V, and repeat steps 4 and 5. Record all results in Table 38-2.
7. Increase the voltage in 0.5 V steps but not beyond that which you calculated in step 2. Note if there are any changes in the lamp's brilliance as V_s is increased.
8. You may be able to measure the voltage across the lamp when S_1 opens by positioning a DMM across the lamp and noting the reading. (An oscilloscope might be more effective at measuring this voltage.)

DATA FOR EXPERIMENT 38

TABLE 38-1 *Inductor parameters*

L	R_{dc}	I_{max}	V_{max}

TABLE 38-2 *Lamp activity*

dc Voltage	S_1 Closing	S_1 Opening

NOTES

QUESTIONS FOR EXPERIMENT 38

( )

1. With S_1 closed, the voltage across the lamp in Figure 38-1 is equal to
   (a) 0 V   **(b)** V_s **(c)** a very high voltage   **(d)** unknown

2. When S_1 is closed, the lamp current is
   (a) essentially zero
   (b) equal to the inductor current

( )   **(c)** very large

3. When S_1 opens, the voltage across the lamp is
   (a) momentarily very high   **(b)** equal to V_s

( )   **(c)** small but no larger than V_s

4. The "kick-back voltage" that causes the lamp to light depends on
   (a) the value of the lamp resistance
   (b) the value of the source voltage
   (c) the value of the dc current flowing through L when S_1 is opened

( )   **(d)** all of the above

5. Explain in your own words why the lamp lights when S_1 is opened. Be sure to include in your discussion the idea of a "collapsing magnetic field."

6. Why was it important not to exceed I_{max} for the inductor?

39

TRANSFORMER FUNDAMENTALS

REFERENCE READING

Principles of Electric Circuits: Sections 14–1 through 14–3.

RELATED PROBLEMS FROM *PRINCIPLES OF ELECTRIC CIRCUITS*

Chapter 14, Problems 3 through 11.

OBJECTIVE

To verify the voltage and turns ratio relationships in a transformer.

EQUIPMENT

Audio signal generator
DMM
Transformer (12.6 V filament type)

BACKGROUND

Transformers, and other devices using the basic principle of mutual inductance, are used in a variety of electronic circuits. Specifically, the transformer is a basic building block in almost all power supplies. Though manufacturers do not always give turns ratios on their data sheets, they can always be deduced from a knowledge of the primary and secondary working voltages. In this experiment, we are going to use the ubiquitous 12.6 V filament transformer, widely available from almost any electrical shop. It is designed to work at 60 Hz with a 115–120 V input and a 12.6 V output. The 12.6 V output is usually the voltage available when the device is operating under a full load. This refers to the maximum secondary current (often 1 A) specified by the manufacturer.

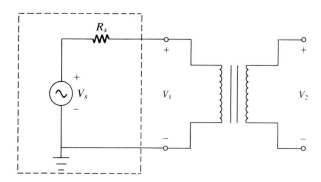

FIGURE 39-1

Consequently, determination of the turns ratio is only approximate using this method. You will, however, use this to calculate expected voltages in the experiment. Don't be too surprised if your predictions are a little off.

PROCEDURE

1. Identify the high- and low-voltage windings of the transformer. In a transformer of this type, the high-voltage (115–120 V) side is normally termed the *primary* and the low-voltage (12.6 V) side, the *secondary.*
2. Use an ac bridge or an ohmmeter to measure the dc resistance of each winding, and record in Table 39-1.
3. Use an ac bridge to measure the self-inductance of each winding, and record in Table 39-1.
4. The transformer is designed to step down the line voltage of 115–120 V to a value of about 12.6 V. To avoid using potentially hazardous voltages, we are going to determine the turns ratio by application of a smaller ac voltage to the primary.
5. Use the *rated* primary and secondary voltage values to determine a theoretical turns ratio for the transformer. Record this in Table 39-2. You should be aware that this will yield a value for the turns ratio under fully loaded conditions and will therefore be less than the apparent ratio for an open-circuited secondary.
6. Set up the circuit in Figure 39-1 with the high-voltage side connected to the generator terminals. Adjust the source to a convenient whole-number rms voltage at a frequency of about 60 Hz.
7. Under Calculated Data in Table 39-2, enter your chosen value from step 6 for V_1. Calculate the expected value of V_2, using this and your theoretical turns ratio.
8. Measure V_2 with your DMM set to ac volts, and record in Table 39-2 under Measured Data.
9. Use the ratio of your measured primary and secondary voltages to calculate the actual turns ratio under no-load, and record in the appropriate column of Table 39-2. How does it compare with the theoretically expected value determined in step 5?
10. Exchange the windings so that what was the secondary now becomes the primary, and what was the primary becomes the secondary.
11. Now repeat steps 5 through 9 and record all data in Table 39-3. Do you see that the transformer now behaves as a step-up type?

Name _____ Date _____

DATA FOR EXPERIMENT 39

TABLE 39-1

	Primary (115–120 V)	Secondary (12.6 V)
Winding Resistance R		
Winding Inductance L		

TABLE 39-2

Calculated Data		Measured Data		Turns Ratio	
V_1	V_2	V_1	V_2	Theoretical	Actual

TABLE 39-3

Calculated Data		Measured Data		Turns Ratio	
V_1	V_2	V_1	V_2	Theoretical	Actual

NOTES

QUESTIONS FOR EXPERIMENT 39

1. In Figure 39-1, if V_1 were 200 mV and V_2 were 4 V, then the turns ratio would be equal to

(   )   (a) 50   (b) 20   (c) 0.05   (d) 0.02

2. When the filament transformer is used in a step-up mode, the turns ratio is closer to

(   )   (a) 1   (b) 0.1   (c) 10   (d) 12.6

3. For a step-up transformer under no-load, which of the following is correct?

(   )   (a) $V_2 < V_1$ (b) $P_2 > P_1$ (c) $I_2 > I_1$ (d) $I_2 = 0$

4. When the transformer is used in the step-down mode the turns ratio is closest to

(   )   (a) 1   (b) 0.1   (c) 10   (d) 12.6

5. The primary inductance is greater than the secondary inductance. Explain why this is so.

6. The primary resistance is greater than the secondary resistance. Explain why this could be so.

40

IMPEDANCE MATCHING

REFERENCE READING

Principles of Electric Circuits: Sections 14–4 through 14–6.

RELATED PROBLEMS FROM *PRINCIPLES OF ELECTRIC CIRCUITS*

Chapter 14, Problems 14 through 19.

OBJECTIVE

To demonstrate impedance matching using a transformer.

EQUIPMENT

Audio signal generator
DMM
Transformer: Any audio frequency impedance-matching transformer, or 12.6 V filament transformer with appropriate source and load terminations
Resistors: Depends on choice of transformer (see Procedure)

BACKGROUND

A transformer can be used to change an impedance level in a circuit to great advantage; in particular, it can transform a load impedance into one of a different value. This is put to use in impedance matching a source of given impedance to a load. In this experiment, you will use a transformer to produce an effective load impedance equal to that of the source. Under these conditions, maximum power is transferred from the source to the load. This experiment is written so that it can be applied to any audio transformer. Once you know the source and load impedances for the par-

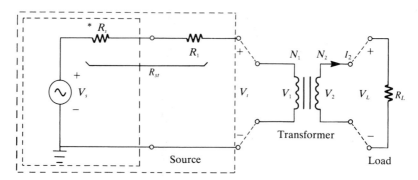

FIGURE 40-1

ticular one you are using, you must "pad"* or supplement the generator resistance to bring its Thevenin resistance to the correct value. This is simply accomplished by adding resistance in series or shunt with the generator.

If you follow the procedure exactly, the experiment will work very nicely. Be sure that you measure open-circuit voltage when instructed. This is one of the few occasions when you set the generator voltage *before* you connect it to the primary coil.

PROCEDURE

1. Determine the values of the source and load impedances your transformer is designed to match. This information should be supplied with the transformer, or consult with your instructor.
2. In this experiment, we shall refer to the source impedance as R_{st} and to the impedance terminating the secondary as the load impedance R_L.
3. Refer to Figure 40-1. The source impedance R_{st} will be equal to the sum of the source resistance of the signal generator (often 600 Ω, but check) and a resistance R_1, which will be inserted to bring the total to R_{st}.
4. Using the value of R_{st} you obtained in step 1, calculate the maximum power available from a source with this value of internal resistance if the open circuit voltage V_s were equal to 4 V. Record your result under Calculated Data in Table 40-1.
5. Suppose now that this power were to be transferred to the secondary load R_L. Using the value of R_L you obtained in step 1, calculate the voltage required to yield this maximum power figure in this load. Record as V_L in Table 40-1.
6. Using the fact that the impedance, as seen looking into the primary, must be equal to the value of R_{st} for maximum power transfer, calculate the voltage that should appear at the primary terminals for open-circuit source voltage of 4 V. Record in the table under V_1.
7. The required turns ratio V_2/V_1 can now be calculated and recorded in Table 40-1 under the n column.
8. Connect the circuit in Figure 40-1. You must build up the source resistance by adding an appropriate value R_1 in series with R_s to bring the total to R_{st} (specified in your transformer data sheet). Do this with discrete resistors or a potentiometer.[†]

---

*"Pad" is a term used that means supplement or compensate for, in electronics.
[†]For example, suppose $R_s = 600\ \Omega$ and $R_{st} = 10\ k\Omega$. Then R_1 should equal $R_{st} - R_s = 9400\ \Omega$.

9. Break the primary circuit, and with a DMM, set up an open circuit voltage of 4 V (measured between the end of R_1 and ground).
10. Reconnect the transformer. Then measure and record V_1 ($= V_t$), and V_2 ($= V_L$). Use the measured value of V_L and Ohm's Law to obtain the secondary current I_L and record in Table 40-2
11. Calculate the power in the load P_L, and record in Table 40-2. Compare these measured data with those obtained theoretically in Table 40-1.

Name _____ Date _____

DATA FOR EXPERIMENT 40

TABLE 40-1 *Calculated data*

P_{max}	V_L	$V_1 = V_t$	$V_2 = V_L$	n

TABLE 40-2 *Measured data*

$V_1 = V_t$	$V_2 = V_L$	I_L	P_L

NOTES

QUESTIONS FOR EXPERIMENT 40

1. If a 4 V source with a 10 kΩ source impedance were connected directly to a 16 Ω load, the load power would be close to

( ) (a) 2.5 μW   (b) 1 W   (c) 250 mW   (d) 0.4 mW

2. In such a case as question 1, the power dissipated in the total source impedance would be close to

( ) (a) 1.6 mW   (b) 0.4 mW   (c) 1.6 W   (d) 96 μW

3. When a transformer matches a source to a load, the voltage appearing at the primary should always be one-half the open-circuit source voltage.

( ) (a) True   (b) False

4. If the transformer were to match a 1000 Ω source to a 16 Ω load, the turns ratio required would be close to

( ) (a) 62.5   (b) 0.016   (c) 2.56×10^{-4} (d) 0.13

5. In a circuit such as that in Figure 40-1, half of the total power is always "wasted." Explain this statement. How efficient is the source when the impedances are matched?

6. The transformer that matches a 10 kΩ source to a 16 Ω load can be used to match other source-load-pair values. What are these, and what is the relationship that must exist between the two?

41

THE SERIES *RC* SINUSOIDAL RESPONSE

REFERENCE READING

Principles of Electric Circuits: Sections 15–2 through 15–4.

RELATED PROBLEMS FROM *PRINCIPLES OF ELECTRIC CIRCUITS*

Chapter 15, Problems 19 through 21, 31, 32, 33.

OBJECTIVE

To examine the frequency response of a series *RC* circuit. To determine the effects of variation in *R*, *C*, and frequency on the voltages and currents.

EQUIPMENT

Audio signal generator
Oscilloscope and two $10\times$ probes
High-impedance voltmeter (optional)
Capacitor ($\pm 10\%$): 0.001 µF
Resistor ($\pm 5\%$): 16 kΩ

BACKGROUND

Because the impedance of a capacitor (its reactance) depends on frequency, any circuit that contains one will have frequency-dependent voltages, currents, and impedances. This fact is put to good use in devices known as filters. We will examine the specific properties of filters in a later experiment. However, it is worth mentioning at this point that the *RC* circuit in Figure 41-1 is, itself, a filter.

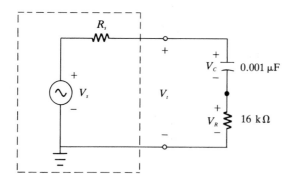

FIGURE 41-1

The determination of how these quantities change as a function of frequency is known as the *frequency response*. In this experiment, we will examine the frequency response of the simple RC circuit. The first part of the experiment will have you make some general observations about the variation of V_R and V_C with frequency. Though you will not record any quantitative data here, it will give you a general "feel" for the overall behavior of this circuit. In part B of the experiment, you will obtain actual frequency response data on the capacitor and resistor voltages. Though there are several formulas you will need to use, and these are adequately explained in the text references, you should keep one very important formula in mind as you perform part B of the experiment. This concerns the fact that the voltages in ac circuits do not, in general, add as they do in dc circuits. In this circuit, for example, you must use the Pythagorean relationship for the voltages:

$$V_t^2 = V_R^2 + V_C^2 \qquad (\text{not } V_t = V_R + V_C)$$

Keeping this in mind as you measure the voltages V_R and V_C in part B will alert you as to whether the results you are getting are sensible.

There are some important considerations to be aware of in performing experiments of this nature. First, the voltage at the generator terminals will not remain constant when the frequency is changed because the load impedance depends on frequency and, as you will recall, the generator has an internal, or "output," resistance. Second, if one side of the signal generator is grounded, placement of instruments is critical to obtain the correct data. In particular, one terminal of voltmeters that are operated from the ac line voltage may be earth-grounded, and therefore must be placed with care in a circuit in which one terminal on the generator is also earth-grounded. This applies also to the oscilloscope, which, of course, is always earth-grounded.

The data in this experiment can be taken with either ac voltmeters or the oscilloscope. If you use ac voltmeters, all data will be in terms of rms values. Some laboratories are not equipped with ac voltmeters that have a high enough impedance or a high enough frequency response for this experiment. In such cases, you should use the oscilloscope for all of your measurements. When using the oscilloscope, then when recording data, peak or peak-to-peak values are the most convenient. If using an oscilloscope, you should use correctly compensated $10\times$ probes, which reduce circuit loading at high frequencies. If you use only $1\times$ (direct) probes, you can expect to get appreciable errors in measurement at the higher frequencies in the tables.

PROCEDURE

Part A: Variation of V_R and V_C with Frequency

1. Measure the actual values of the resistor and capacitor, and record these in the Notes section of the data sheet. You will need these later when verifying calculations.
2. Construct the circuit shown in Figure 41-1. The resistor R_s inside the dotted outline is the signal generator's internal resistance and is shown to remind you that it is always present and affects the terminal voltage V_t when the frequency is altered.
3. With channel 1 connected to the signal generator, and channel 2 to the resistor voltage, set the generator for a sine-wave output at a frequency of 10 kHz and any convenient amplitude. (Larger amplitudes yield better results.)
4. Set up the oscilloscope so that you can see the terminal voltage and resistor voltage simultaneously (dual-trace mode).
5. The resistor voltage should be somewhat smaller than the terminal voltage and should lead it by about one-eighth of a cycle ($\approx 45°$) at this frequency.
6. Vary the frequency above and below 10 kHz, and observe the resistor voltage. Its amplitude should change with frequency. A change in frequency of one decade (a factor of ten) above and below 10 kHz is normally adequate for you to be able to see these variations. Record your observations in Table 41-1, using the arrow symbology: ↑ means increase, ↓ means decrease, and → means no change.
7. Now examine the phase angle θ between V_R and V_t. Because the circuit current is in phase with V_R, the angle between these two traces represents the impedance angle. By varying the frequency around 10 kHz, you should be able to see how the magnitude of the angle |θ| changes with frequency. Use the arrow symbology to enter these data in Table 41-1.
8. To observe the capacitor voltage V_C, *you must generally switch (physically exchange) the positions of R and C so that the capacitor is grounded and the resistor is not.* Another method involves using the **ADD/INVERT** feature provided on most oscilloscopes. Using this feature, you can view the capacitor voltage without physically moving any of the components in the circuit. Your instructor can tell you more about this feature.
9. Using either of the methods described in step 8, determine the direction in which the capacitor voltage changes with frequency. Once again, a decade above and below 10 kHz should be enough to make the variation apparent.

Part B: Frequency Response Data

In this part of the experiment, if using voltmeters for the measurement of data, assume rms quantities for the word *amplitude* in the following steps. If using the oscilloscope with $10\times$ probes, assume peak values when the word *amplitude* is used. In either case, have your instructor make sure that all instruments are capable of making good measurements to 50 kHz in this circuit.

1. Restore the circuit to its original form (with the resistor grounded).
2. If using the oscilloscope for your measurements, be sure to use a $10\times$ probe and, before you make any measurements, carefully *frequency compensate* both probes. Your instructor will show you how to compensate a $10\times$ probe.
3. Set up the generator for an amplitude of 2 V at 2 kHz.

4. Measure the amplitude of the voltage across the resistor at this frequency and record in Table 41-2.

5. By exchanging the components, using the oscilloscope **ADD/INVERT** mode, or otherwise, determine the amplitude of the capacitor voltage at this frequency. Before you go any further, make a quick calculation to see if the (phasor) sum of these voltages gives approximately 2 V (the terminal voltage) within about 10 percent. If not, check your circuit and measurement technique before you proceed.

6. For each frequency in Table 41-2, measure and record the resistor and capacitor voltages, recording all data in Table 41-2. *Be sure to maintain the terminal voltage constant at 2 V as you change the frequency.* Depending on the output resistance of the generator, this voltage might change appreciably, especially as you move to higher frequencies.

7. Use the measured data and the actual (measured) values of R and C to complete the table. Formulas are given for each calculation.

8. For each frequency in Table 41-3, calculate the quantities X_C, Z, I, V_R, and V_C. Once again, use the actual component values if possible. These calculations should compare reasonably well with the corresponding measured data from Table 41-2.

DATA FOR EXPERIMENT 41

TABLE 41-1   *Variational effects*

Quantity Varied	Effects					
$f \uparrow$	V_R		V_C		$\lvert\theta\rvert$	
$f \downarrow$	V_R		V_C		$\lvert\theta\rvert$	

TABLE 41-2   *Measured data*

Frequency	V_R	V_C	$I\left(=\dfrac{V_R}{R}\right)$	$X_C\left(=\dfrac{V_C}{I}\right)$	$Z\left(=\dfrac{V_t}{I}\right)$
2 kHz					
5 kHz					
10 kHz					
20 kHz					
50 kHz					

TABLE 41-3   *Calculated data*

Frequency	$X_C\left(=\dfrac{1}{\omega C}\right)$	$Z(=\sqrt{R^2 + X_C^2})$	$I\left(=\dfrac{V_t}{Z}\right)$	$V_R(=IR)$	$V_C(=IX_C)$
2 kHz					
5 kHz					
10 kHz					
20 kHz					
50 kHz					

NOTES

Name ——————————————————— Date ——————————

QUESTIONS FOR EXPERIMENT 41

( )

1. In a series RC circuit such as this, the maximum possible value of $|\theta|$ is
    (a) 90°   **(b)** 180°   **(c)** 360°   **(d)** 45°

2. In a series RC circuit such as this, as the frequency increases, the current
    (a) decreases   **(b)** increases   **(c)** remains constant

( )

    (d) sometimes increases and sometimes decreases

3. In this circuit, as the frequency increases, the capacitor voltage V_C
    (a) decreases   **(b)** increases   **(c)** stays constant

( )

    (d) increases, then decreases

4. Refer to Figure 41-1. At a frequency of approximately 10 kHz, the voltages across R and C are each one-half of the terminal voltage.

( )

    (a) True   **(b)** False

5. Draw a phasor diagram for this circuit, with $V_t = 2$ V, showing all voltages at a frequency of 10 kHz. Use a current reference. What is the relationship between the resistor voltage, the capacitor voltage, and the generator voltage at this frequency?

6. Why is it necessary to maintain a constant generator voltage when performing a frequency response? What would happen to the frequency response data if the generator voltage were allowed to vary?

230

42

THE PARALLEL *RC* SINUSOIDAL RESPONSE

REFERENCE READING

Principles of Electric Circuits: Sections 15–5 through 15–6.

RELATED PROBLEMS FROM *PRINCIPLES OF ELECTRIC CIRCUITS*

Chapter 15, Problems 40, 43.

OBJECTIVE

To examine the frequency response of a parallel *RC* circuit. To determine the effects of variation in *R, C,* and frequency on the branch currents.

EQUIPMENT

Audio signal generator
Oscilloscope and two 10× probes
High-impedance voltmeter
Capacitor (±10%): 0.001 μF
Resistors (±5%): 16 kΩ
　　　　three 100 Ω precision

BACKGROUND

The last experiment investigated the series *RC* circuit response. The primary quantities of interest were the capacitor and resistor voltages. When a resistor and capacitor are connected in a parallel arrangement, the voltages across each are equal, and it is the variation of their currents and the total current that is of interest. Instruments that will measure small ac currents accurately and without appreciable loading effects are difficult to make and are, consequently, often absent from electronics laboratories. This is easily overcome by using a current-monitoring tech-

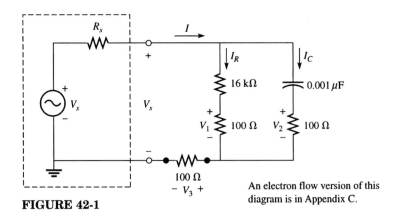

FIGURE 42-1

An electron flow version of this diagram is in Appendix C.

nique. The 100 Ω resistors in Figure 42-1 are for this purpose. Because the current in each branch must flow through the respective resistor, a voltage measurement across each will allow us to calculate the magnitude of this current. Notice that the resistors are small in value. Why do you think this is the case? A further application of this technique is in the measurement of the total current *I*.

Once again, be aware of the problems caused by "multiple grounds." Connect all grounded instruments to the same ground as that of the signal generator. (See experiment 41 Background.)

Finally, the data in this experiment can be taken with either ac voltmeters or the oscilloscope. Some laboratories are not equipped with ac voltmeters that have a high enough impedance or a high enough frequency response for this experiment. In such cases, you should use the oscilloscope for all of your measurements. If using an oscilloscope, you should use $10\times$ probes, which reduce circuit loading at high frequencies. If you use only $1\times$ (direct) probes, you can expect to get appreciable errors in measurement at the higher frequencies.

PROCEDURE

Part A: Variation of I_R and I_C with Frequency

1. Measure the actual values of the resistors and capacitor, and record these in the Notes section on the data sheet. You will need these later when verifying calculations.
2. Construct the circuit shown in Figure 42-1, omitting, for the moment, the 100 Ω resistor in series with the source at the bottom of the schematic. The resistor R_s inside the dashed outline is the signal generator's internal resistance and is shown to remind you that it is always present and affects the terminal voltage when the frequency is altered.
3. With channel 1 connected to the loaded signal generator, set it for a sine-wave output (V_t) at a frequency of 10 kHz and any convenient amplitude.
4. Move the channel 1 probe and connect it so that it monitors V_1 in the figure. Connect channel 2 to monitor V_2. Set up the oscilloscope so that you can see both of these voltages simultaneously (dual-trace mode).
5. At 10 kHz, the voltages should be approximately equal in amplitude but will be out of phase with each other by about one-quarter of a cycle ($\approx 90°$).
6. Vary the frequency above and below 10 kHz, and observe any variation in amplitude of either of these two voltages. Pay attention to the phase angle between these voltages. The amplitude of one should change, while that of the other should remain essentially constant. A change in frequency of just a few kHz is normally adequate for you to be able to see these variations. Recall that the

current I_R is proportional to V_1, and the current I_c is proportional to V_2. With this in mind, record your observations in Table 42-1 using the arrow symbology: ↑ means increase, ↓ means decrease, and → means no change.

7. Restore the frequency to 10 kHz. Insert the third 100 Ω resistor, in series with the source as shown. Remove the channel 1 probe from the circuit, and use the channel 2 probe to measure the voltage V_3 at the bottom of the schematic. (Remember that both the channel 1 and channel 2 probe ground leads must *never* be connected to electrically different points in the same circuit.) This voltage is proportional to the total current I in the circuit.

8. Once again, by varying the frequency around 10 kHz, determine the direction in which this voltage, and hence I, changes with frequency.

Part B: Frequency Response Data

In this part of the experiment, use the oscilloscope for all measurements. Assume peak values when the word *amplitude* is used. Additionally, you should use frequency-compensated $10 \times$ probes in this experiment.

1. Measure the values of the three 100 Ω resistors, as well as that of the 16 kΩ resistor, and record these in the Notes section of the data sheet. Make sure you know *which* 100 Ω resistor is in *which* position in the circuit. You will need this information in steps 4 and 9. If possible, measure the capacitance of the capacitor and record its value in the Notes section.

2. Be sure to use $10 \times$ probes with the oscilloscope and, before you make any measurements, carefully compensate both probes. Your instructor will show you how to compensate a $10 \times$ probe.

3. Set up the generator for an amplitude of at least 2 V at 2 kHz. Omit the "bottom" 100 Ω resistor for the moment.

4. Leave the channel 1 probe connected across the generator terminals, and use the channel 2 probe to separately measure the amplitudes of the voltages V_1 and V_2 at this frequency. Record these amplitudes in Table 42-2. Use these measured values together with the actual values of the 100 Ω resistors to compute the currents I_R and I_C through the resistive and capacitive branches.

5. Now insert the other 100 Ω resistor, remove the channel 1 probe from the circuit, and use the channel 2 probe to measure the voltage V_3. (Remember that both the channel 1 and channel 2 probe ground leads must *never* be connected to electrically different points in the same circuit.)

6. Using the measured voltage for V_3 and the actual value of the 100 Ω resistor, calculate the total current I in the circuit. Before you go any further, make a quick calculation to see if the (phasor) sum of the two branch currents, I_R and I_C, gives approximately the total current I. If not, then check your circuit and measurement technique before you proceed.

7. Next, for each frequency in Table 42-2, measure and record the three individual resistor voltages. *Be sure to maintain the terminal voltage constant at 2 V as you change the frequency.* Depending on the output resistance of the generator, this voltage might change appreciably, especially as you move to the higher frequencies. This is the reason for keeping the channel 1 probe across the generator terminals and moving only the channel 2 probe.

8. Use the measured data to complete the table. Formulas are given for each calculation.

9. For each frequency in Table 42-3, calculate the quantities X_C, I_R, I_C, I, and Z. Assume that this is an ideal parallel RC circuit; that is, ignore the 100 Ω resistors in the calculations. Once again, use the actual values of R and C if possible. These calculations should compare reasonably well with the corresponding measured data from Table 42-2.

DATA FOR EXPERIMENT 42

TABLE 42-1   *Variational effects*

Quantity Varied	Effects					
$f \uparrow$	I_R		I_C		I	
$f \downarrow$	I_R		I_C		I	

TABLE 42-2   *Measured data*

Frequency	V_1	V_2	V_3	I_R	I_C	I	$I\,(= \sqrt{I_R^2 + I_C^2})$	$Z\,(= V_t/I)$	$X_C + (= V_t/I_C)$
2 kHz									
5 kHz									
10 kHz									
20 kHz									

TABLE 42-3   *Calculated data*

Frequency	$X_C \left(= \dfrac{1}{\omega C}\right)$	$I_R \left(= \dfrac{V_t}{R}\right)$	$I_C \left(= \dfrac{V_t}{X_C}\right)$	$I\,(= \sqrt{I_R^2 + I_C^2})$	$Z \left(= \dfrac{V_t}{I}\right)$
2 kHz					
5 kHz					
10 kHz					
20 kHz					

NOTES

QUESTIONS FOR EXPERIMENT 42

1. In Table 42-2, as the frequency was increased, the branch current I_R

( ) (a) increased   (b) decreased   (c) remained constant

2. In Table 42-2, as the frequency was decreased, the branch current I_C

( ) (a) increased   (b) decreased   (c) remained constant

3. In this circuit, the phase angle between the branch currents I_R and I_C
   (a) increases with frequency
   (b) decreases with frequency
   (c) does not change with frequency

( ) (d) increases, then decreases, with frequency

4. At a frequency of approximately 10 kHz, the total current is

( ) (a) 0   (b) $2 I_C$ (c) $2 I_R$ (d) $\sqrt{2 I_R}$

5. Draw a phasor diagram for this circuit, showing all currents at a frequency of 10 kHz. Use a terminal voltage reference phasor. What is the approximate relationship between the branch currents and the generator current at this frequency?

6. Explain the reason for making the current-monitoring resistors very small in a circuit such as this. What is the disadvantage in making them too small?

43

THE SERIES *RL* SINUSOIDAL RESPONSE

REFERENCE READING

Principles of Electric Circuits: Sections 16–1 through 16–3.

RELATED PROBLEMS FROM *PRINCIPLES OF ELECTRIC CIRCUITS*

Chapter 16, Problems 1, 2, 3, 5, 6, 10, 12, 13, 14, 15.

OBJECTIVE

To observe the relationships between impedance, current, inductive voltage, and resistive voltage in an *RL* ac circuit.

EQUIPMENT

Audio signal generator
Oscilloscope and two 10× probes
DMM
Inductor: 1–5 H (ac resistance < 500 Ω at 1 kHz)
Resistor (±5%): 33 kΩ

BACKGROUND

Circuits containing inductance and resistance appear in a variety of electronic circuits, from power supplies to filters. In this experiment we are going to investigate the sinusoidal response of a series *RL* circuit. A difficulty arises in conjunction with such circuits in that real inductors are not the ideal inductances we deal with in our theory. Since they are formed of coiled wire, they possess resistance as well as inductance. Furthermore, their resistance is dependent on frequency. As a consequence, the inserted resistance *R* does not represent the total resistance of the circuit. In addition, when we measure the voltage across a coil, we are getting both

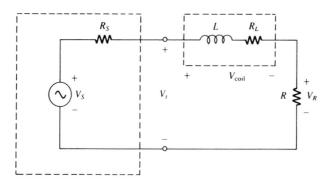

FIGURE 43-1

inductive and resistive components of voltage, not simply V_L. In this experiment, we will try to overcome this problem by making R large compared with the ac resistance of the coil; that is, we will presume the coil is ideal. Do not be surprised if the results of this experiment are a little off from your expectations.

PROCEDURE

In this experiment, if using voltmeters for the measurement of data, assume rms quantities for the word *amplitude* in the following steps. If using the oscilloscope with $10\times$ probes, assume peak values when the word *amplitude* is used.

1. Determine the actual inductance and dc resistance of the coil. The dc resistance can be found using the DMM on **OHMS** and the inductance on an impedance bridge or meter. Measure and record the actual value of the 33 kΩ resistor. Record these measured values in Table 43-1.
2. Using the actual values of the circuit components, complete the table for the calculated values, assuming a sine wave of amplitude 5 V and frequency of 1 kHz.
3. Connect the circuit in Figure 43-1, and adjust the terminal voltage of the generator for 5 V at 1 kHz.
4. Measure the coil voltage V_{coil} and resistor voltage V_R, recording the measured data in Table 43-2. If using grounded instruments, remember to exchange the positions of the inductor and resistor when measuring the inductor voltage (or else use the **ADD/INVERT** mode on the oscilloscope).
5. Use the measured values to complete the table. If the coil's ac resistance at 1 kHz is small with respect to 33 kΩ, the data should compare reasonably well with those in Table 43-1.
6. You can repeat all the above measurements at another frequency determined by your instructor. Record all data in Tables 43-3 and 43-4.

DATA FOR EXPERIMENT 43

TABLE 43-1

Measured Coil Parameters			Calculated Circuit Quantities at 1 kHz						
L	R_L	R	Z_{coil}	Z	I	V_R	V_L	V_{R_L}	V_{coil}

TABLE 43-2

V_R	V_{coil}	$I = \dfrac{V_R}{R}$	$Z_{\text{coil}} = \dfrac{V_{\text{coil}}}{I}$	$Z = \dfrac{V_t}{I}$

TABLE 43-3

Calculated Quantities at ( ) kHz						
Z_{coil}	Z	I	V_R	V_L	V_{R_L}	V_{coil}

TABLE 43-4

V_R	V_{coil}	$I = \dfrac{V_R}{R}$	$Z_{\text{coil}} = \dfrac{V_{\text{coil}}}{I}$	$Z = \dfrac{V_t}{I}$

NOTES

QUESTIONS FOR EXPERIMENT 43

1. If $L = 5$ H and $R_L = 300\ \Omega$, the expected value of I in Figure 43-1 for a terminal voltage of 5 V at 1 kHz is
    (a) 0.943 mA   (b) 0.109 mA
( )   (c) 10.159 mA   (d) 0.998 mA

2. If the resistance of the coil in question 1 were ignored, the expected value for the current would be
    (a) 159 mA   (b) 0.159 mA
( )   (c) 1.00 mA   (d) 0.109 mA

3. Based on your answers to questions 1 and 2, you might conclude that in this case
    (a) coil resistance is insignificant and can be ignored
( )   (b) coil resistance is significant and must be taken into account

4. If the coil resistance is negligible, the phase difference between V_{coil} and V_R
    (a) is always 90°
    (b) is equal to 90° at only one frequency
    (c) is equal to 45°
( )   (d) is directly proportional to the frequency

5. Explain any discrepancies between the expected value of V_{coil} and that measured in step 4 of the procedure.

6. Draw a phasor diagram of the calculated voltages in Figure 43-1. Include I as a reference phasor, and show the positions of V_{R_L}, V_L, V_{coil}, and V_t relative to the current.

44

THE SERIES *RLC* CIRCUIT

REFERENCE READING

Principles of Electric Circuits: Sections 17–1 and 17–2.

RELATED PROBLEMS FROM *PRINCIPLES OF ELECTRIC CIRCUITS*

Chapter 17, Problems 1 through 6.

OBJECTIVE

To examine voltage current and impedance relationships in a series *RLC* circuit.

EQUIPMENT

Audio signal generator
Oscilloscope and two $10\times$ probes
High-impedance voltmeter
Capacitor ($\pm5\%$): 0.47 μF*
Inductor: 100–200 mH radio frequency coil
Resistor ($\pm5\%$): 1 kΩ

BACKGROUND

Series *RLC* circuits are considerably more complicated to analyze than those with only *L* and *R* or *C* and *R*. When a circuit contains both inductance and capacitance, the possibility of antiphase† voltages arises, and unless one is aware of the effects of these, measured voltages may seem obscure or erroneous.

---

*This experiment works best if *C* has a value such that $X_C \cong X_L/2$ at 1 kHz.
†Antiphase is a term used to describe a voltage that is 180° out of phase with another voltage.

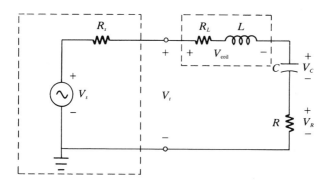

FIGURE 44-1

A further major difference in *RLC* circuits is that they may behave inductive, capacitive, or resistive, depending on the applied frequency. In this experiment, we will take measurements at one frequency only and reserve a complete frequency response for a later experiment.

When measuring the voltages V_R, V_{coil}, and V_C, remember to use a *floating* meter; otherwise exchange the physical positions of the components to facilitate voltage measurements with grounded meters. Remember also that the coil is not an ideal inductance, and V_{coil} contains both resistive and inductive components of voltage.

PROCEDURE

In this experiment, if using voltmeters for the measurement of data, assume rms quantities for the word *amplitude* in the following steps. If using the oscilloscope with $10\times$ probes, assume peak values when the word *amplitude* is used. Note: A higher voltage than 2 V may be used.

1. Measure the actual values of all components, and use these measured values to calculate the quantities in Table 44-1 at a frequency of 1 kHz. Assume a generator terminal voltage amplitude of 2 V. Assume the coil is ideal; that is, has an ac resistance of 0 Ω for these calculations. *NOTE:* $X_T = X_L - X_C$, $\theta = \tan^{-1} (X_T/R)$.
2. Construct the circuit in Figure 44-1 with a generator amplitude of 2 V and frequency of 1 kHz.
3. Measure the voltages V_{coil}, V_C, and V_R. If using grounded measuring instruments, be sure that you exchange the positions of the components correctly. Record data in Table 44-2.
4. The phase angle θ represents the phase difference between the terminal voltage V_t and the source current I. Because the voltage V_R is in phase with the current I, this voltage also represents the phase of the current. Connect the oscilloscope with channel 1 on V_t and channel 2 on V_R. Set it up for the dual-trace mode and trigger from channel 1.
5. With both channels on a simultaneous ground reference—say, in the center of the screen—use the method outlined in experiment 32 to measure the phase difference between the displayed voltages. You will recall that this involves *spreading* the waveform so that one cycle fits into exactly eight major divisions

across the screen. Then, using the fact that one division represents 45°, measure the phase angle between the channel 1 and channel 2 signals. Record this in Table 44-2.

6. Complete the table by calculating the quantities in the remaining columns.

7. Using the measured values, complete the labeling in Tables 44-3 and 44-4 for the voltage and impedance triangles.

Name _____ Date _____

DATA FOR EXPERIMENT 44

TABLE 44-1 *Calculated values*

X_L	X_C	$Z = \sqrt{R^2 + X_T^2}$	$I = V_t/Z$	V_L	V_C	V_R	θ

TABLE 44-2 *Measured values*

V_{coil}	V_C	V_R	$I = V_R/R$	$X_C = V_C/I$	$Z = V_t/I$	θ

TABLE 44-3

 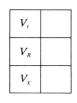

V_t	
V_R	
V_X	

TABLE 44-4

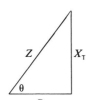

Z	
R	
X_T	

NOTES

QUESTIONS FOR EXPERIMENT 44

1. In Figure 44-1, if $V_{coil} = 7$ V, $V_C = 3$ V, and V_R is 3 V, then you would expect V_t to be
( )   (a) 13 V   (b) 8.18 V   (c) 10.4 V   (d) 5 V

2. In Figure 44-1, if $V_R = 2.6$ V, $V_C = 1.25$ V, and R is 1 kΩ, then X_C has a value of
( )   (a) 1 kΩ (b) 481 Ω (c) 208 Ω (d) 339 Ω

3. In Figure 44-1, if $V_C = 1.25$ V and $V_{coil} = 2.5$ V, then the *net* voltage across an ideal coil and capacitor ought to be close to
( )   (a) 3.75 V   (b) 2.8 V   (c) 1.25 V   (d) 0

4. For a given current, a real coil when compared with an *ideal* coil will have
(a) the same voltage
(b) a greater voltage
( )   (c) a smaller voltage

5. What effect would the ac resistance of the coil have on the impedance triangle in Table 44-4? In particular, how will the sides of the triangle be affected?

6. Show with the aid of sketches how a series RLC circuit can look either inductive or capacitive, depending on the magnitude of X_L and X_C.

45

SERIES RESONANCE

REFERENCE READING

Principles of Electric Circuits: Section 17–3.

RELATED PROBLEMS FROM *PRINCIPLES OF ELECTRIC CIRCUITS*

Chapter 17, Problems 10 through 14.

OBJECTIVE

To show that the resonant frequency of a series *RLC* circuit is given by $1/(2\pi\sqrt{LC})$.
To plot the frequency response of an *RLC* circuit.

EQUIPMENT

Audio signal generator
Oscilloscope and two $10\times$ probes
Capacitors: 0.001 μF and 0.01 μF
Resistors ($\pm 5\%$): 100 Ω (two)
Inductor: 100–200 mH radio frequency coil

BACKGROUND

We examined the voltage relationships in a series *RLC* circuit in experiment 44. Now we are ready to see how these voltages and other quantities vary with frequency.

In the first part of the experiment, we will use the voltages across *R, L,* and *C* to locate the resonant frequency and note how they change around resonance. Since you will be using the oscilloscope for this, the component of interest has to be positioned with one end grounded to avoid "grounding out" components.

The second part of the experiment involves the variation of V_R with frequency. This, through Ohm's Law, is also the variation of *I* with frequency. The circuit must

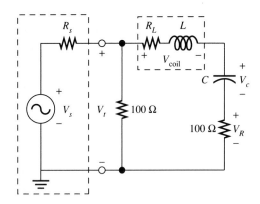

FIGURE 45-1

be fed with a constant voltage as usual, and you will have to maintain this condition throughout the frequency range of the experiment. This brings us to the purpose of the 100 Ω resistor that is shown across the generator terminals (Figure 45-1). Because the impedance falls to a minimum at the resonant frequency, the source voltage V_s will divide between R_s and the load impedance, which will be small. This causes V_t to change radically as the condition of resonance is approached. To minimize this, the 100 Ω resistor is placed across the terminals. This effectively reduces the Thevenin-equivalent output impedance of the generator to a value less than 100 Ω. Because it is closer to an ideal source now, the load will not have such a pronounced effect on V_t around the resonant frequency. This resistor will, of course, reduce the amount of voltage available at the terminals, and if this is a problem, it can be removed with no detrimental effect except for the above. The 100 Ω resistor has no effect whatsoever in the calculations pertaining to the load; consider it simply as part of the signal generator (as though it were 'invisible' to you).

PROCEDURE

If using the oscilloscope with $10\times$ probes, assume peak values when the word *amplitude* is used. Have your instructor make sure that all instruments are capable of making good measurements at 100 kHz in this circuit.

1. Use the *measured* values of the inductance and each of the capacitors to determine the series resonant frequency of the circuit in Figure 45-1. Record these calculated values in Table 45-1.
2. Construct the circuit with $C = 0.001$ μF. Unless your instructor specifically instructs you otherwise, be sure to connect the additional 100 Ω resistor directly across the terminals of the function generator.
3. With its output connected to channel 1 of the oscilloscope, adjust the generator for a peak amplitude of 4 V at 1 kHz. If the generator is not able to go to 4 V (because of the 100 Ω resistor across its terminals), use a smaller voltage that is convenient, such as 1 or 2 V.
4. Trigger from channel 1, and connect channel 2 to monitor the resistor voltage V_R. Set up the oscilloscope for dual-trace operation.
5. Vary the frequency in a direction that produces an *increasing* voltage across the resistor. Measure and record, in Table 45-1, the frequency that makes this voltage a maximum. This is the resonant frequency of the circuit.
6. Substitute the 0.01 μF capacitor and repeat step 5.

7. Leave this larger capacitor in the circuit, and adjust the frequency to a value below resonance. This corresponds to the symbol $f < f_r$ in the first two rows of Table 45-2. Now vary the frequency a little on either side of this frequency, and observe and record the direction in which V_R changes. Record these variations in the first two rows of Table 45-2, using the arrow notation. Be sure not to increase the frequency so far that you go through resonance—*stay below f_r.*

8. Now change the frequency so that you are above resonance. This corresponds to the symbol $f > f_r$ in the second two rows of Table 45-2. Once again, vary the frequency a little on either side of this frequency, and observe and record the direction in which V_R changes. Record these variations in the second two rows of Table 45-2. Be sure not to decrease the frequency so far that you go through resonance—*stay above f_r.*

9. Restore the frequency to a value below resonance. While varying the frequency you should be able to see the phase of V_R relative to V_t changing. Because the phase of V_R is the same as that of the circuit current, this angle represents the magnitude of the impedance angle of the circuit. The magnitude of this phase difference is denoted by $|\theta|$ in the table. By varying the frequency and staying *below f_r,* record the variation of $|\theta|$ in the first two rows of the table.

10. Repeat step 9 at a frequency above resonance and record the data in the second two rows of the table.

11. Exchange the positions of R and C, and, with channel 2 now monitoring the capacitor voltage, determine the variation in V_C above and below resonance as you did with V_R.

12. Now exchange the positions of L and C, and repeat this procedure for V_L, recording the data in the last column of Table 45-2.

13. Restore R to the grounded position in the circuit (the relative positions of L and C are not important), and set the generator frequency to 100 Hz.

14. Adjust the generator voltage to a convenient value such as 1 V peak.

15. Turn your attention to Table 45-3. In it, you see a listing of frequencies beginning at 100 Hz and extending through 100 kHz. Because the dials of many generators are not well calibrated in terms of frequency, you should use the oscilloscope to verify the accuracy of the frequency setting. Except for f_r, which you have noted earlier, these frequencies increase in a 1-2-5 sequence (e.g., 100 Hz, 200 Hz, 500 Hz, etc.). This sequence is particularly easy to set up using the oscilloscope because the reciprocals of these numbers, which represent the period, form a 1-5-2 sequence (i.e., 10 ms, 5 ms, 2 ms, etc.), which corresponds to the **SECS/DIV** sequence on most oscilloscopes.

16. For each frequency listed in Table 45-3, measure and record the resistor voltage while keeping the generator terminal voltage constant at your chosen value from step 14. *Be sure that you constantly monitor the generator's terminal voltage being displayed on channel 1 of the oscilloscope because its value is likely to change, especially around resonance.*

17. For each frequency in Table 45-4, use the true values of L, C, and R to calculate the quantities X_C, X_L, Z, I, and V_R. The values you obtain for V_R will have the same general form as those you measured, but will differ in magnitude because these calculations ignore the ac coil resistance.

18. On the next pages, graph paper is provided for you to plot both the measured (Table 45-3) and the theoretical (Table 45-4) values for V_R. The frequency assignments on the "f axis" are already made for you and are in logarithmic form. You need to decide only on the scale for the V_R axis. Plot graphs of practical and theoretical data.

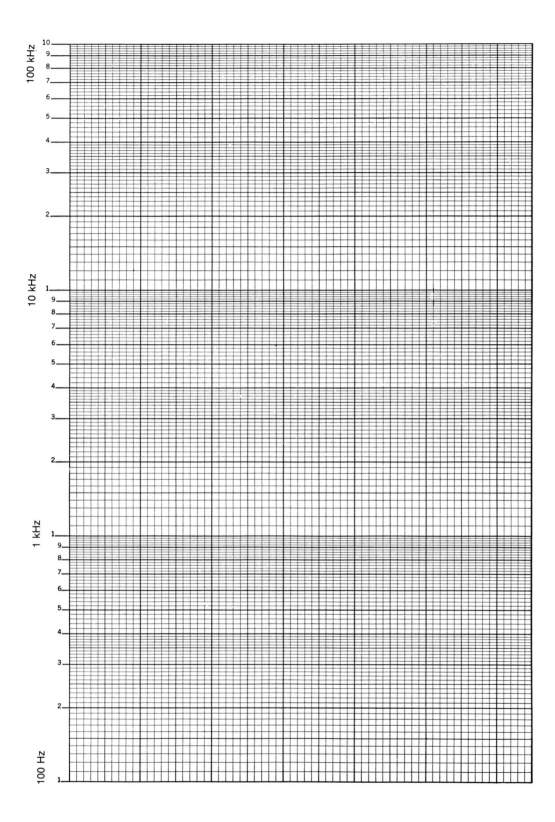

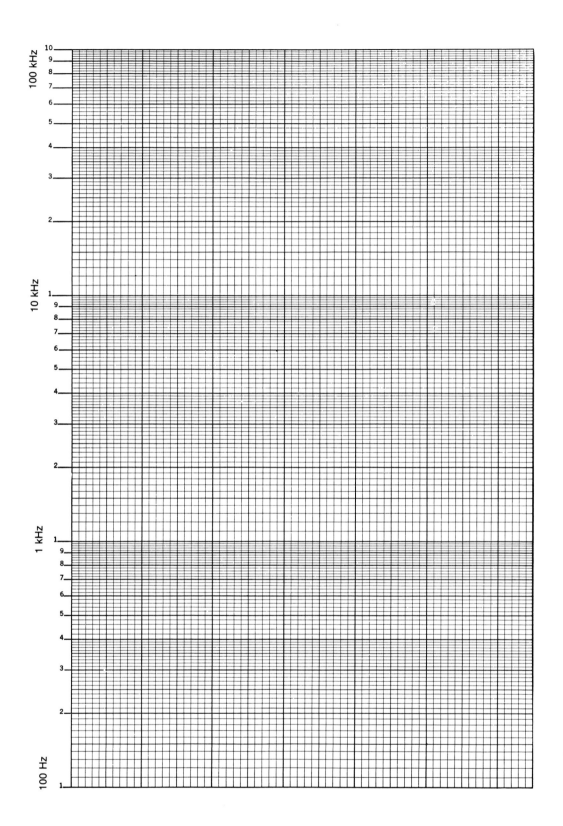

Name _____   Date _____

DATA FOR EXPERIMENT 45

TABLE 45-1

Measured Component Values		Resonant Frequency	
L	C	Calculated	Measured
	0.001 μF		
	0.01 μF		

TABLE 45-2   *Variational effects*

Quantity Varied	Effects						
$f < f_r : f \uparrow$	V_R		$|\theta|$		V_C		V_L	
$f < f_r : f \downarrow$	V_R		$|\theta|$		V_C		V_L	
$f > f_r : f \uparrow$	V_R		$|\theta|$		V_C		V_L	
$f > f_r : f \downarrow$	V_R		$|\theta|$		V_C		V_L	

TABLE 45-3   *Measured data*

Frequency	V_R
100 Hz	
200 Hz	
500 Hz	
1 kHz	
2 kHz	
f_r	
5 kHz	
10 kHz	
20 kHz	
50 kHz	
100 kHz	

Name _____ Date _____

TABLE 45-4 *Calculated data*

Frequency	$X_C\left(=\dfrac{1}{\omega C}\right)$	$X_L(=\omega L)$	$Z(=\sqrt{R^2+[X_L-X_C]^2}\,)$	$I\left(=\dfrac{V_t}{Z}\right)$	$V_R(=IR)$
100 Hz					
200 Hz					
500 kHz					
1 kHz					
2 kHz					
f_r					
5 kHz					
10 kHz					
20 kHz					
50 kHz					
100 kHz					

NOTES

QUESTIONS FOR EXPERIMENT 45

( )
1. With $L = 100$ mH and $C = 0.001$ μF, the resonant frequency is
   (a) 1000 Hz **(b)** 10 GHz **(c)** 100 kHz **(d)** 15.9 kHz

( )
2. In a series circuit such as this, as the frequency increases the current always decreases.
   (a) True **(b)** False

3. In this circuit, if the frequency is below that of resonance, and *increases,* then
   (a) V_R increases while Z increases
   (b) V_R increases while Z decreases
   (c) V_R increases while Z remains constant
( )
   (d) V_R decreases while Z increases

4. In this circuit, as you increase the frequency from *very low* through to *very high* values,
   (a) the current increases
   (b) the current decreases
   (c) the current first increases, then decreases
( )
   (d) the current first decreases, then increases

5. In performing a frequency response as in this experiment, why is it necessary to maintain a constant input voltage?

6. Comment on the effect that ac coil resistance has on a circuit such as this. Pay particular attention to its effect at the resonant frequency.

46

PARALLEL RESONANCE

REFERENCE READING

Principles of Electric Circuits: Sections 17–4 through 17–6.

RELATED PROBLEMS FROM *PRINCIPLES OF ELECTRIC CIRCUITS*

Chapter 17, Problems 22 through 25.

OBJECTIVE

To determine the frequency characteristics of a parallel resonant circuit.

EQUIPMENT

Audio signal generator
Oscilloscope and two $10 \times$ probes
Capacitor: 0.001 μF
Resistors ($\pm$ 5%): 1 MΩ, 100 Ω
Inductor: 100–200 mH radio frequency coil

BACKGROUND

The parallel *LC* circuit, or *tank circuit* as it is often called, has a frequency response similar in many ways to that of the series *RLC* circuit. When a constant-amplitude voltage is applied to an *LC* circuit, and the frequency increased from a low value, the impedance increases to a maximum at resonance and then falls off as the frequency increases still further. If the coil resistance is small (or the coil *Q* is ≥10), the resonant frequency is approximately the same as it is when the coil and capacitor are connected in series; i.e., $f_r = 1/(2\pi\sqrt{LC})$. The impedance of the circuit is infinite. When coil resistance is *not* negligibly small, the resonant frequency formula is modified (see Reference Reading), and the impedance at resonance is given by L/CR_{coil}.

In this experiment we are going to investigate the variation of input current to a near-ideal parallel LC circuit when the input voltage amplitude is kept constant. This is the objective of part A. The resistor R in Figure 46-1 is used as a current monitoring resistor; that is, its voltage allows us to calculate the input current to the tank circuit.

Because tank circuits are often driven by constant-current sources, part B investigates how the tank circuit voltage varies when the input current amplitude is maintained constant. To create an approximate constant-current source, a large resistor is inserted in series with the signal generator. So long as this resistor is much larger than the impedance of the tank circuit at any frequency, the input current will be approximately constant even though the frequency changes. The 1 MΩ resistor is large enough so that the tank circuit will receive an input current that is approximately constant. The tank circuit voltage and impedance can then be plotted against frequency. *As in some of the previous experiments, the frequencies in this experiment are high enough that compensated 10× probes must be used for accuracy.*

PROCEDURE

In this experiment, if using the oscilloscope with frequency-compensated 10× probes, assume peak values when the word *amplitude* is used. Have your instructor make sure that all instruments are capable of making good measurements at 100 kHz in this circuit.

Part A: Constant-Voltage Input—Current Response

1. Using an impedance bridge, or other instrument, measure and record the actual values of *L, C,* and *R* for the 100 Ω resistor, and record the data in Table 46-1.
2. Using the actual measured values for *L* and *C,* calculate and record the expected parallel resonant frequency. Treat the coil as ideal (zero ac resistance) for the purpose of these calculations.
3. With its output connected to channel 1 of the oscilloscope, set the function generator for an output of 10 V p–p at a frequency of 10 kHz.
4. Construct the circuit in Figure 46-1, and, with channel 1 monitoring the generator's terminal voltage, and channel 2 on the resistor voltage, vary the frequency until the resistor voltage is a *minimum*. Record this as the measured frequency of resonance in Table 46-1.
5. Set the generator frequency to 1 kHz, and adjust the generator terminal voltage to some convenient value such as 1, 2, or 4 V p–p. (Use the largest convenient p–p voltage to improve your accuracy.)
6. For each frequency in Table 46-2, measure and record the resistor voltage V_R while maintaining V_t constant at your chosen value in step 5. Be sure to keep a watchful eye on the terminal voltage as it will tend to vary with frequency. Use p–p voltages and currents in the tables.
7. Using each measured voltage, and the measured value of *R* from step 1, calculate and record the circuit total current *I* for each frequency in the table.
8. Plot a graph of *I* p–p versus log frequency on the graph paper provided.

Part B: Constant-Current Input—Voltage Response

1. Using an impedance bridge, or other instrument, measure the actual values of *L, C,* and the 1 MΩ resistor, and record the data in Table 46-3.

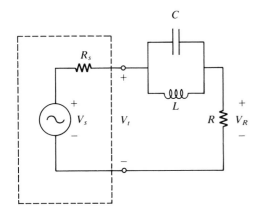

FIGURE 46-1

2. Using the actual measured values for L and C, calculate and record the expected parallel resonant frequency. Treat the coil as ideal (zero ac resistance) for the purpose of these calculations.

3. Construct the circuit in Figure 46-2, and set the generator's terminal voltage for an output of 10 V p–p at a frequency of 10 kHz.

4. With channel 1 monitoring the generator's terminal voltage, and channel 2 monitoring the parallel-tuned circuit voltage V_o, vary the frequency until V_o is a maximum. Record this frequency as the measured frequency of resonance in Table 46-3.

5. Set the generator frequency to 10 kHz, and adjust the generator terminal voltage to 10 V p–p. (If 10 V p–p is not available, use the largest convenient p–p voltage to improve accuracy.)

6. Vary the frequency from about 1 kHz to 100 kHz and observe the generator terminal voltage V_t. It should not appreciably vary over this range. If this is the case, then you can assume that the input current does not vary much (this means that the voltage drop across the internal resistance of the generator is staying fairly constant). The circuit is therefore being driven by an *approximate constant-current source*.

7. Calculate and record in Table 46-4 the magnitude of this constant current using $I_{p-p} = V_{p-p}/R$ where V_{p-p} is the peak-to-peak input voltage, from step 5, and R is the measured resistance from step 1. (We are here assuming that the impedance of the parallel LC circuit is always much less than 1 MΩ.)

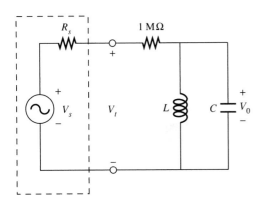

FIGURE 46-2

8. For each frequency in Table 46-4, measure and record the voltage V_O *while maintaining V_t constant at your chosen value in step 5.* The frequencies listed are easy to measure with the oscilloscope because of the 1-2-5 sequence. However, if you have a frequency counter, or if the frequency calibration on your generator allows, you can take data at intermediate frequencies; this will improve the quality of the graphs in step 10.

9. Record the voltage V_o at the resonant frequency.

10. For each of the frequencies, using the current I_{p-p} from step 7 and voltage V_o, calculate and record the impedance Z_p of the parallel LC circuit.

11. Plot graphs of V_o and Z_p versus log frequency on the graph paper provided.

DATA FOR EXPERIMENT 46

TABLE 46-1

Measured Component Values			Resonant Frequency	
R	L	C	Calculated	Measured

TABLE 46-2

Frequency (kHz)	V_R p–p (volts)	Ip–p (mA)
1		
2		
5		
10		
20		
50		
100		

TABLE 46-3

Measured Component Values			Resonant Frequency	
L	R	C	Calculated	Measured

TABLE 46-4

Input Current (μA p–p)		
Frequency (kHz)	V_o p–p (volts)	Z_p (kΩ)
1		
2		
5		
10		
20		
50		
100		
200		
Measured Resonant Frequency (from Table 46–3)		

NOTES

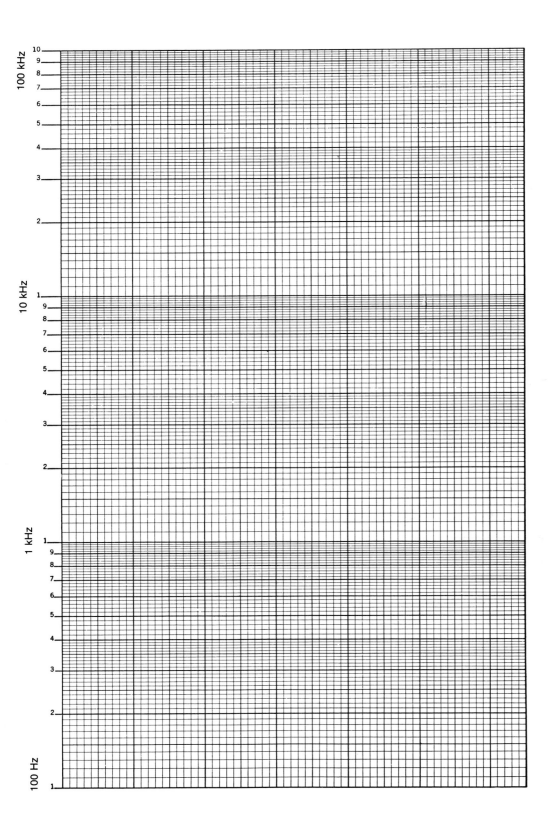

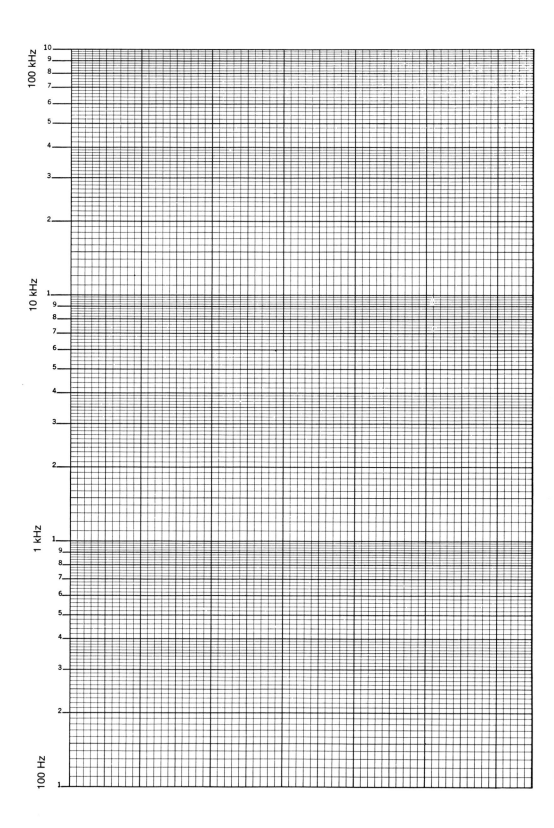

QUESTIONS FOR EXPERIMENT 46

( )

1. With $L = 100$ mH, and $C = 0.001$ μF, the circuit's resonant frequency is
   (a) 1000 Hz   (b) 10 GHz   (c) 100 kHz   (d) 15.9 kHz

2. In Figure 46-1, at resonance,
   (a) the voltage V_R is a maximum
   (b) The voltage V_o is a maximum

( )

3. In Figure 46-2, when R is much greater than the impedance of the parallel LC circuit, then the circuit is said to be driven by
   (a) an approximate constant voltage
   (b) an approximate constant current
   (c) neither of these

( )

4. In a parallel LC circuit, as the frequency increases from very low to very high values, the impedance of the circuit
   (a) increases   (b) decreases
   (c) increases, then decreases   (d) decreases, then increases

( )

5. Describe the effect that coil resistance (which you ignored) has on this circuit's resonant frequency.

6. The impedance of a parallel LC circuit at resonance is given by L/CR_{coil}. Use the data that you obtained in part B of this experiment about the circuit's impedance at resonance to predict the inductor's coil resistance.

47

LOW-PASS FILTERS (LAG NETWORKS)

REFERENCE READING

Principles of Electric Circuits: Section 18–1.

RELATED PROBLEMS FROM *PRINCIPLES OF ELECTRIC CIRCUITS*

Chapter 18, Problems 1 through 10.

OBJECTIVE

To determine the frequency characteristics of low-pass filters.

EQUIPMENT

Audio signal generator
Oscilloscope and $10 \times$ probe
High-impedance millivoltmeter
Capacitor: 0.01 μF
Resistor ($\pm 5\%$): 16 kΩ

BACKGROUND

The simple *RC* circuit we first examined in experiment 41 can be used as an electrical filter. A low-pass filter tends to pass "low" frequencies and attenuate (stop) "high" frequencies. The words "low" and "high" are within quotation marks because the boundaries between these depend upon the component values so that what is high for one circuit may be low for another.

To act as a low-pass filter, the *RC* circuit is set up so that the output voltage is taken across *C* as in Figure 47-1. The circuit is often referred to as a *lag network* in this configuration, because the output voltage V_{out} lags the input voltage V_{in}.

In order that different filter circuits may be compared, the ratio $V_{\text{out}}/V_{\text{in}}$ is calculated or measured as a function of frequency. The ratio is often called *voltage gain*

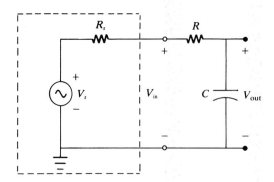

FIGURE 47-1

or *voltage transfer ratio* and serves to tell us whether a signal of given frequency will emerge from the network with little or much attenuation. The ratio is often expressed in dB, from which a Bode plot can be constructed (see Reference Reading).

Of particular interest is the critical frequency f_c, that frequency which causes the output voltage to be 0.707 of its maximum value in this case. For the RC circuit, this is given by $f_c = 1/(2\pi RC)$. It is also the frequency when the output voltage lags the input voltage by 45°.

As with all frequency response measurements, be sure to maintain V_{in} constant for each data point. The semilog graph paper is provided so that you can easily plot $\log f$ without actually taking the log of frequency. Just use it as you would a linear scale.

PROCEDURE

If you are using high-impedance meters for the voltage measurements in this experiment, assume amplitude values are rms. If an oscilloscope is being used for measurement, assume amplitude values are peak. Be sure to compensate any $10\times$ probes that you are going to use in this experiment; otherwise your measurements will be invalid. Note: Voltages higher than 2 V can be used.

1. Measure the actual values of R and C, and use these to calculate the expected critical (or cutoff) frequency f_c. Record this in Table 47-1.
2. Construct the circuit, and, with the oscilloscope in the dual-trace mode, monitor the filter's input voltage (same as the generator's output voltage) on Channel 1 and the output voltage (across the capacitor) on Channel 2.
3. At a frequency of 100 Hz, adjust the filter's input voltage for an amplitude of 2 V. If you are working with voltmeters, treat this as an rms amplitude; otherwise, if exclusively using the oscilloscope, treat it as a peak amplitude.
4. At this frequency, the output voltage of the filter should be very close in amplitude to the input voltage, and the phase difference (lag) should be almost zero.
5. Measure, and record in Table 47-2, the output voltage of the filter.
6. Use the standard method developed in experiment 34 to measure the phase difference between the input and the output voltages, and record in the table. Because the output voltage *lags* the input voltage, the phase angle will be negative. At 100 Hz, this phase difference will be almost zero degrees and therefore difficult to measure accurately.
7. Next, for each frequency in Table 47-2, measure and record both the magnitude and phase of the output voltage relative to the input, while maintaining V_{in}

constant at 2 V. As the frequency increases, the output voltage should decrease, and lag V_{in} by an angle that approaches 90°. Do not complete the Gain and dB rows at this time.

8. Locate, and record in Table 47-1, the critical frequency of the filter. You can do this by either looking for the frequency at which the output voltage has fallen to 0.71×2 V $= 1.42$ V, or using the fact that the phase difference between the input and output voltages is 45°, or one-eighth of a cycle at this frequency. (If you are using high-impedance voltmeters with dB scales, these are particularly convenient for such measurements, and your instructor will demonstrate them.)

9. Complete the Gain (V_{out}/V_{in}) and dB rows of Table 47-2.

10. For each frequency in Table 47-3, calculate the quantities X_C, Z, I, Φ, V_{out}, V_{out}/V_{in}, and V_{out}/V_{in} (dB) for a V_{in} of 2 V. Use the actual (measured) values for R and C; otherwise use the nominal values. Formulas are provided in the tables.

11. Graph paper is provided to plot the voltage gain, either directly or in dB, and the phase angle Φ versus (log) frequency for your measured and calculated data—four plots in all. See your instructor for the plots actually required.

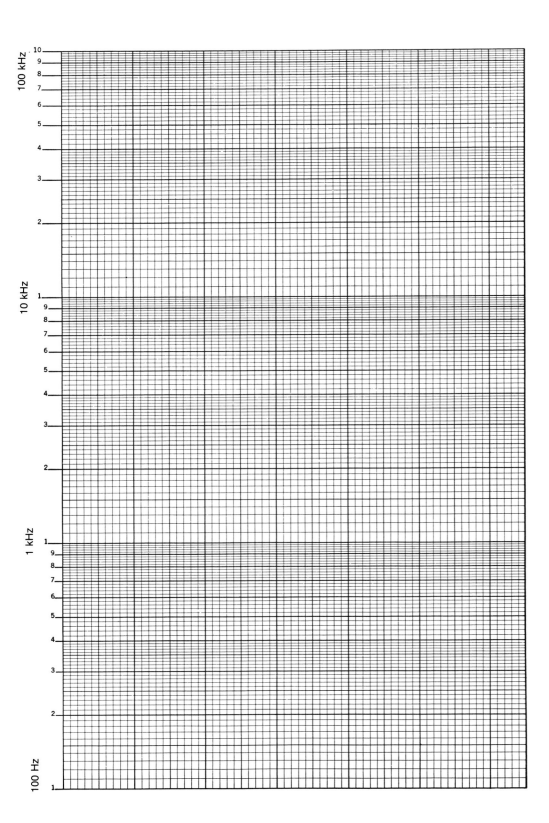

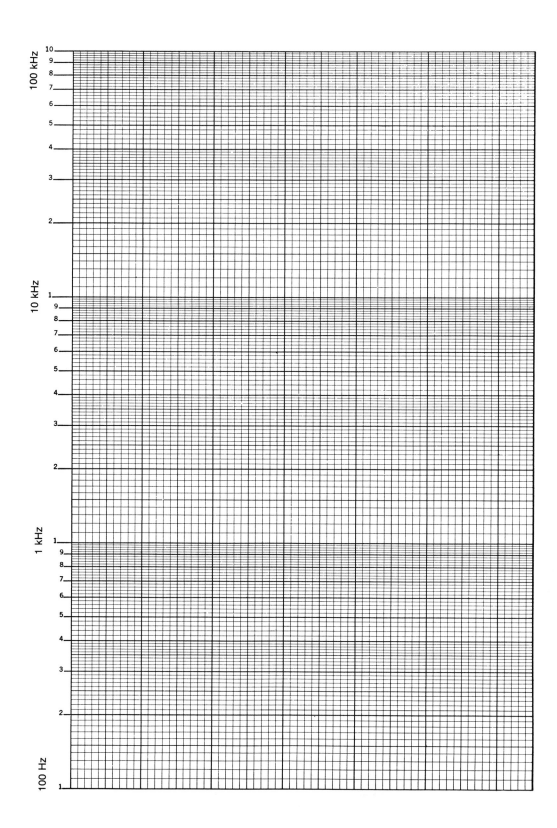

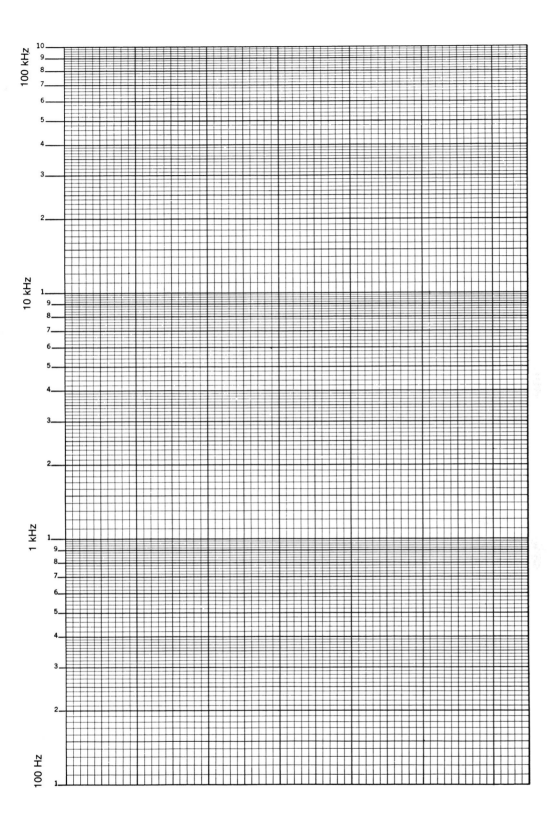

100 kHz

10 kHz

1 kHz

100 Hz

Name _____ Date _____

DATA FOR EXPERIMENT 47

TABLE 47-1

Measured Component Values		Critical Frequency f_c	
R	C	Calculated	Measured

TABLE 47-2 *Frequency response*

Frequency Response				$V_{in} =$						
Frequency (kHz)	0.1	0.2	0.5	1	2	5	10	20	50	100
Output Voltage (V)										
Phase Φ (°)										
Gain $= V_{out}/V_{in}$										
20 log V_{out}/V_{in} (dB)										

TABLE 47-3   *Frequency response*

Frequency (kHz)	0.1	0.2	0.5	1	2	5	10	20	50	100
$X_C = \dfrac{1}{\omega C}$										
$Z = \sqrt{R^2 + X_C^2}$										
$I = \dfrac{V_t}{Z}$										
Phase $\Phi = \tan^{-1}\dfrac{X_C}{R}$										
$V_{\text{out}} = IX_C$										
$V_{\text{out}}/V_{\text{in}}$										
$20 \log V_{\text{out}}/V_{\text{in}}$ (dB)										

NOTES

QUESTIONS FOR EXPERIMENT 47

    1. With values as in Figure 47-1, the voltage gain ratio V_{out}/V_{in} at a frequency of 10 kHz is closer to

(   )   **(a)** 10   **(b)** 1   **(c)** 0.1   **(d)** 0.01

    2. In the low-pass filter, the output voltage

        (a) lags the input voltage

        (b) leads the input voltage

(   )   **(c)** is in phase with the input voltage

    3. In the low-pass filter, the output voltage does not change very much during the interval where

(   )   **(a)** f is much less than f_c **(b)** f is much greater than f_c

    4. At the critical frequency f_c, the phase difference between V_{out} and V_{in} was close to

(   )   **(a)** $30°$ **(b)** $-45°$ **(c)** $-90°$ **(d)** $0°$

    5. Explain the difficulties that would arise if log scales for frequency were not used.

    6. Can you think of a practical use for a low-pass filter? (*Hint:* Audio equipment often has one.)

48

HIGH-PASS FILTERS (LEAD NETWORKS)

REFERENCE READING

Principles of Electric Circuits: Section 18–2.

RELATED PROBLEMS FROM *PRINCIPLES OF ELECTRIC CIRCUITS*

Chapter 18, Problems 11 through 15.

OBJECTIVE

To determine the frequency characteristics of high-pass filters.

EQUIPMENT

Audio signal generator
Oscilloscope and $10 \times$ probe
High-impedance millivoltmeter
Capacitor: 0.001 μF
Resistor ($\pm 5\%$): 16 kΩ

BACKGROUND

In this experiment we are going to once again examine the *RC* circuit, but this time we will take V_{out} across the resistor. The ratio V_{out}/V_{in} will now represent the response of a high-pass filter. Because V_{out} will lead V_{in} at all frequencies, the circuit is called a *lead network*. You should find the response curves of the low-pass filter and high-pass filter to be mirror images of one another. One last comment on the high-pass filter: the response will not remain "flat" as the input frequency increases, but will probably decrease at some high value. You will be asked to consider the reason for this at the end of the experiment.

279

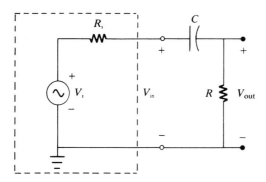

FIGURE 48-1

PROCEDURE

If you are using high-impedance meters for the voltage measurements in this experiment, assume amplitude values are rms. If an oscilloscope is being used for measurement, assume amplitude values are peak. Be sure to compensate any $10\times$ probes that you are going to use in this experiment; otherwise your measurements will be invalid. Note: Voltages higher than 2 V can be used.

1. Measure the actual values of R and C, and use these to calculate the expected critical (or cutoff) frequency f_c. Record this in Table 48-1.
2. Construct the circuit shown in Figure 48-1, and, with the oscilloscope in the dual-trace mode, monitor the filter's input voltage (same as the generator's output voltage) on channel 1 and the output voltage (across the resistor) on channel 2.
3. At a frequency of 100 Hz, adjust the filter's input voltage for an amplitude of 2 V. If you are working with voltmeters, treat this as an rms amplitude; otherwise, if exclusively using the oscilloscope, treat it as a peak amplitude.
4. At this frequency, the output voltage of the filter should be *very small* in amplitude, and the phase difference (lead) should be almost 90°.
5. Measure, and record in Table 48-2, the output voltage of the filter.
6. Use the standard method developed in experiment 34 to measure the phase difference between the input and output voltages, and record in the table. Because the output voltage *leads* the input voltage, the phase angle will be positive.
7. For each frequency in Table 48-2, measure and record both the magnitude and phase of the output voltage relative to the input, while maintaining V_{in} constant at 2 V. As the frequency increases, the output voltage should increase and lead V_{in} by an angle that approaches zero degrees. Do not complete the Gain and dB rows at this time.
8. Locate, and record in Table 48-1, the critical frequency of the filter. You can do this by either looking for the frequency at which the output voltage has fallen to $0.71 \times 2\,\text{V} = 1.42\,\text{V}$ or using the fact that the phase difference between the input and output voltages is 45°, or one-eighth of a cycle at this frequency. (If you are using high-impedance voltmeters with dB scales, these are particularly convenient for such measurements, and your instructor will demonstrate them.)
9. Complete the Gain (V_{out}/V_{in}) and dB rows of Table 48-2.
10. For each frequency in Table 48-3, calculate the quantities X_C, Z, I, Φ, V_{out}, V_{out}/V_{in}, and V_{out}/V_{in} (dB) for a V_{in} of 2 V. Use the actual (measured) values for R and C; otherwise use the nominal values. Formulas are provided in the tables.
11. Graph paper is provided to plot the voltage gain, either directly or in dB, and the phase angle Φ versus (log) frequency for your measured and calculated data—four plots in all. See your instructor for the plots actually required.

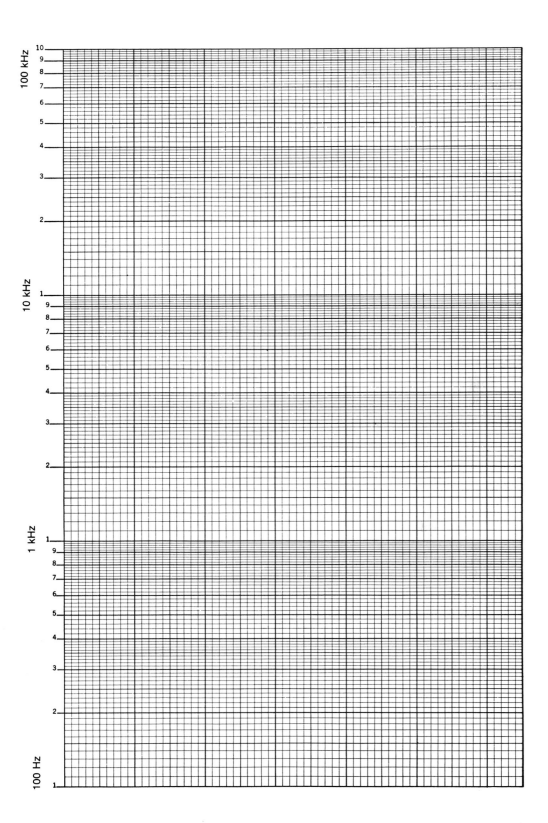

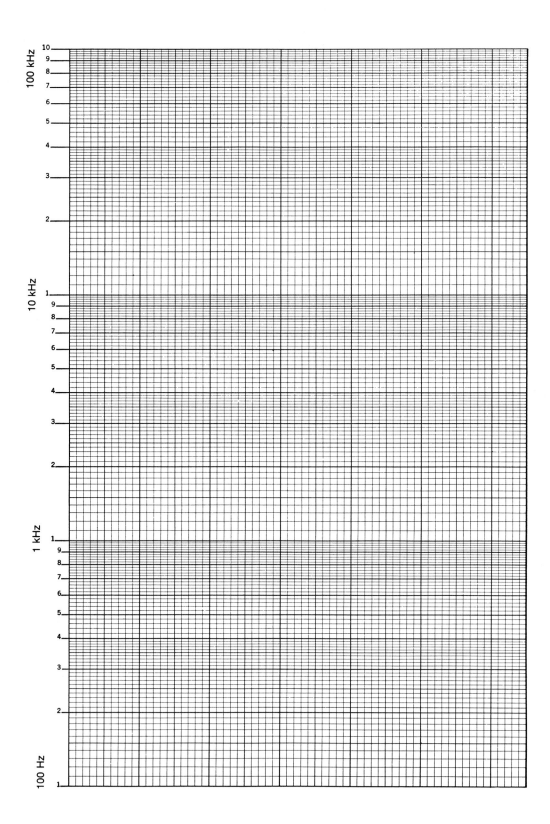

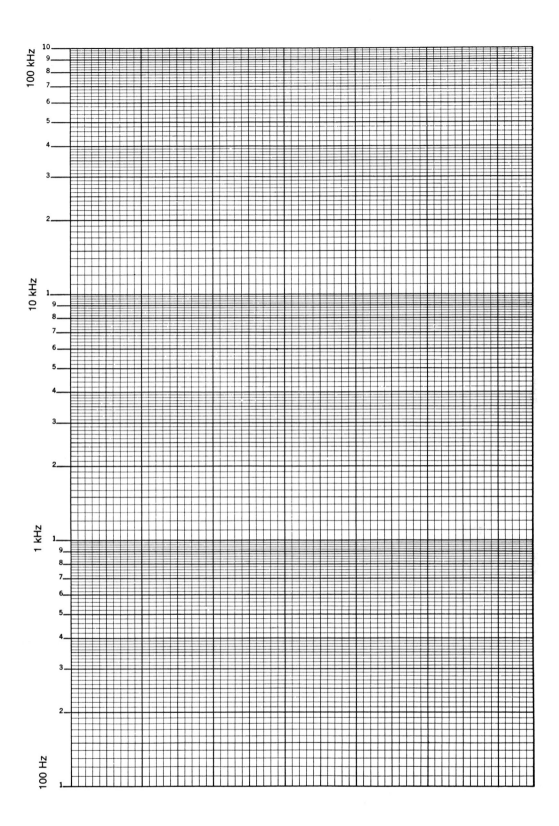

Name _____ Date _____

DATA FOR EXPERIMENT 48

TABLE 48-1

Component Values		Critical Frequency f_c	
R	C	Calculated	Measured
16 kΩ	0.001 μF		

TABLE 48-2 *Frequency response*

Frequency (kHz)	0.1	0.2	0.5	1	2	5	10	20	50	100
Output Voltage (V)										
Phase Φ										
Gain $= V_{out}/V_{in}$										
20 log V_{out}/V_{in} (dB)										

TABLE 48-3 *Frequency response*

Frequency (kHz)	0.1	0.2	0.5	1	2	5	10	20	50	100
X_C										
Z										
I										
Phase $\Phi = \tan^{-1} \dfrac{X_C}{R}$										
V_{out}										
V_{out}/V_{in}										
20 log V_{out}/V_{in} (dB)										

NOTES

QUESTIONS FOR EXPERIMENT 48

1. With values as in Figure 48-1, the voltage gain ratio V_{out}/V_{in} at a frequency of 1 kHz is about

(   )   (a) 10   (b) 1   (c) 0.1   (d) 0.01

2. In the high-pass filter, the output voltage
   (a) lags the input voltage
   (b) leads the input voltage

(   )   (c) is in-phase with the input voltage

3. In the high-pass filter, the output voltage does not change very much during the interval where

(   )   (a) f is much less than f_c (b) f is much greater than f_c

4. At the critical frequency f_c, the phase difference between V_{out} and V_t is close to

(   )   (a) $-30°$ (b) $45°$ (c) $90°$ (d) $0°$

5. The rate at which this filter "rolls off" at low frequencies is said to be 20 dB/decade. Can you explain the meaning of this statement in terms of the way the output voltage varies with frequency at frequencies below f_c?

6. Can you think of a practical use for a high-pass filter? (*Hint:* Audio equipment often has one.)

49

BAND-PASS FILTERS

REFERENCE READING

Principles of Electric Circuits: Section 18–3.

RELATED PROBLEMS FROM *PRINCIPLES OF ELECTRIC CIRCUITS*

Chapter 18, Problems 17 through 21.

OBJECTIVE

To determine the frequency characteristics of series-resonant band-pass filters.

EQUIPMENT

Audio signal generator
Oscilloscope and $10\times$ probe
High-impedance millivoltmeter
Capacitors: 0.01 μF, 0.001 μF
Inductor: 100–200 mH radio frequency coil
Resistors ($\pm 5\%$): 1 kΩ
　　　　　　　　　　 2 kΩ

BACKGROUND

The band-pass filter combines the responses of the low- and high-pass types, resulting, in this case, in a filter that tends to pass a band of frequencies around a central value and attenuate low and high frequencies relative to this center. The series-resonant circuit produces a response like this, the center frequency being the frequency of resonance. In this experiment, we will see how the center frequently can be moved by using different values of capacitance with the same value of inductance. The steepness of the response skirts (see Reference Reading) can be changed by adjusting the circuit resistance R. This in turn changes the bandwidth and selectivity Q, as defined by the equation $BW = f_r/Q$ where f_r is the resonant frequency.

As always, constancy of V_{in} is of key importance for a valid response curve. Finally, recall that the total circuit resistance includes that of the coil (which is not constant with frequency). This will have an effect on your results. You are asked to consider this in the questions at the end of the experiment.

PROCEDURE

If you are using high-impedance meters for the voltage measurements in this experiment, assume amplitude values are rms. If an oscilloscope is being used for measurement, assume amplitude values are peak. Be sure to compensate any $10\times$ probes that you are going to use in this experiment; otherwise your measurements will be invalid.

1. Measure the actual values of R, L, and each of the capacitors, and use these to calculate the expected resonant frequency f_r, the circuit Q, and the bandwidth BW in each row of Table 49-1. Insert the measured values of R and C in the appropriate blank spaces below the corresponding nominal values. Insert the value of L under the L heading. In these calculations, note that you are ignoring the ac resistance of the coil. For better accuracy, include the dc coil resistance in the calculation for the circuit Q. You can easily determine this with an ohmmeter.

2. Construct the circuit shown in Figure 49-1, with values as in the first row of Table 49-1, and, with the oscilloscope in the dual-trace mode, monitor the filter's input voltage (same as the generator's output voltage) on channel 1 and the output voltage (across the resistor) on channel 2.

3. At a frequency of 500 Hz, adjust the filter's input voltage for an amplitude of 2 V. If you are working with voltmeters, treat this as an rms amplitude; otherwise, if exclusively using the oscilloscope, treat it as a peak amplitude. Be consistent.

4. At this frequency, the output voltage of the filter should be *quite small* in amplitude, and it ought to lead the input voltage by some angle.

5. Measure, and record in Table 49-2, the output voltage of the filter at this frequency.

6. Locate and record, under the Measured heading in Table 49-1, the resonant frequency of the filter. You can do this by looking for the frequency at which the output voltage reaches a maximum or by using the fact that the phase difference between the input and output voltages is 0° at this frequency. Record also V_{out} at f_r in Table 49-2.

7. Now adjust the frequency of the generator for 1 kHz, and then, for each of the remaining frequencies in Table 49-2, measure and record the magnitude of the output voltage relative to the input, **while maintaining V_{in} constant at 2 V.** Watch this carefully because the generator's output voltage will tend to decrease as you approach the circuit's resonant frequency. As the frequency increases, the resistor voltage V_R should at first increase and then decrease as you pass over the resonant frequency.

8. Complete the Gain (V_{out}/V_{in}) and dB rows of Table 49-2.

9. Repeat the frequency response for reach of the remaining pairs of resistor and capacitor values in Table 49-1, recording pertinent data in this table and in Tables 49-2 and 49-3.

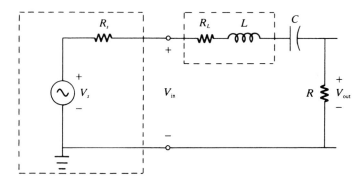

FIGURE 49-1

10. Graph paper is provided to plot the voltage gain either directly or in dB versus (log) frequency for your measured data. See your instructor for the plots actually required.

11. Depending on the plots you make, the actual bandwidth and Q factor of each circuit you built can be determined from them. To determine the bandwidth, measure (on the plots) the distance between the frequencies, on either side of resonance, at which the output voltage has fallen to $0.707 \times V_{max}$, where V_{max} is the maximum (resonant) resistor voltage. The actual circuit Q is determined by $Q = f_r/BW$, as stated in the Background section. You can compare these with the calculated data that ignored the ac coil resistance.

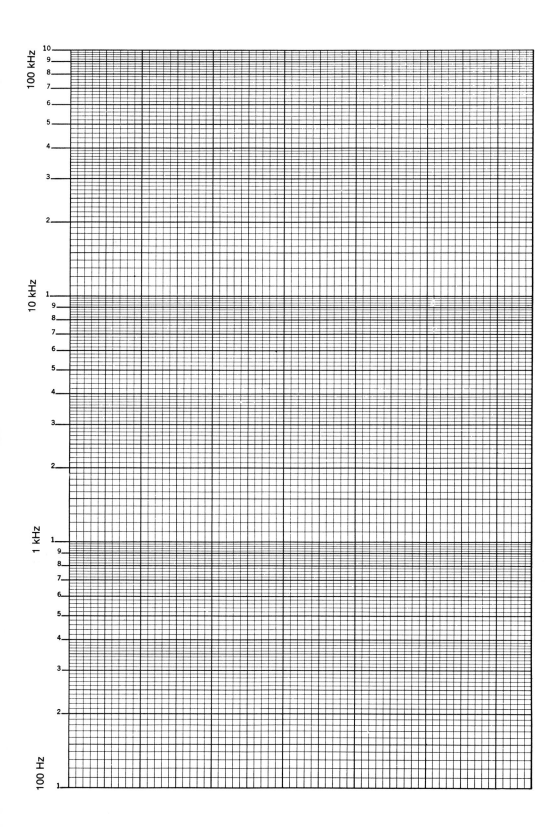

100 kHz

10 kHz

1 kHz

100 Hz

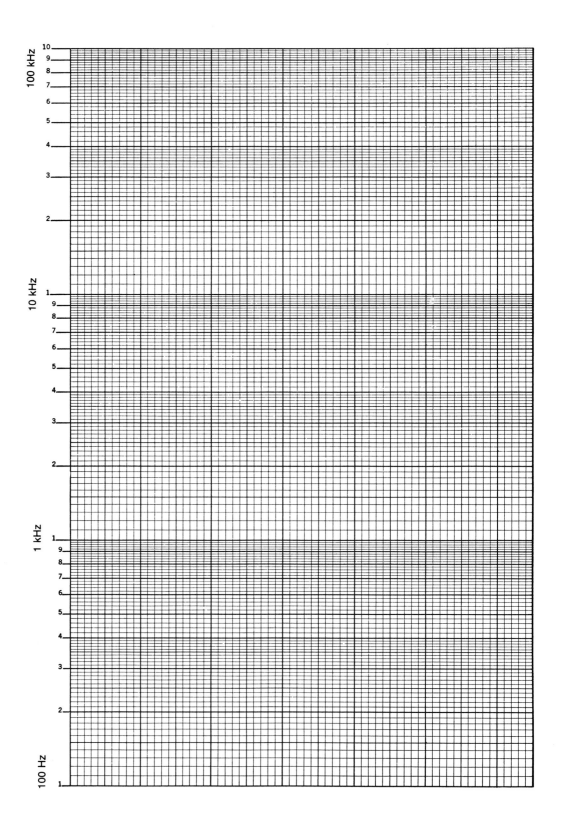

292

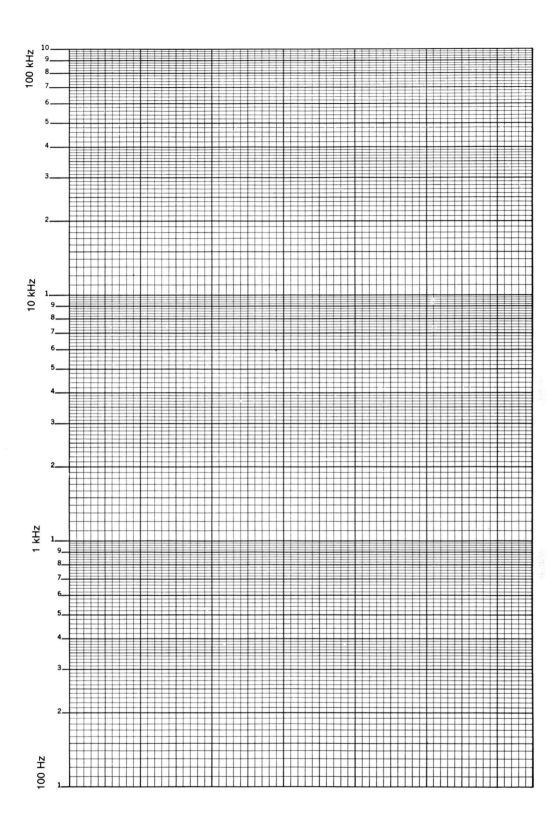

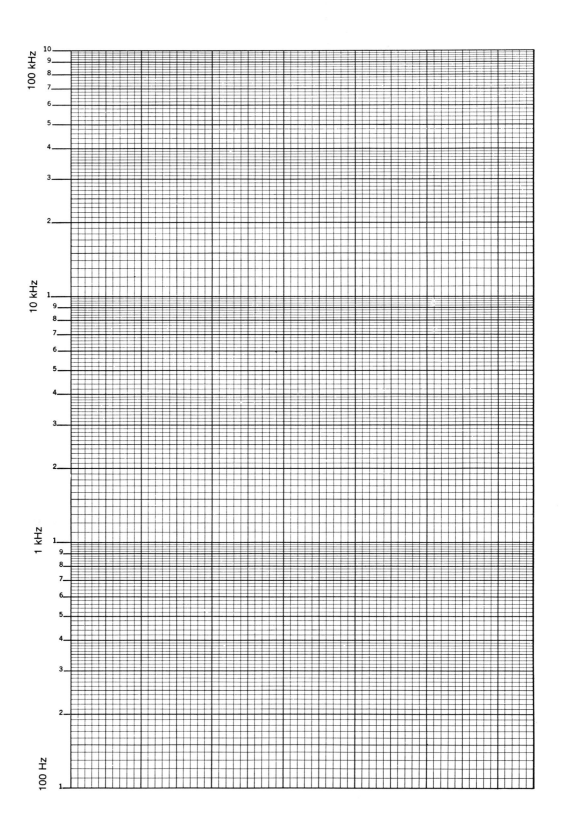

DATA FOR EXPERIMENT 49

TABLE 49-1

Component Values			Resonant Frequency f_r		Q		BW	
	L	C	R	Calculated	Measured	Calculated	Measured	Calculated	Measured
1		0.01 μF	1 kΩ						
2		0.01 μF	2 kΩ						
3		0.001 μF	1 kΩ						
4		0.001 μF	2 kΩ						

TABLE 49-2

$C = 0.01\ \mu\text{F}$		Frequency Response: $V_{in} =$								
Frequency (kHz)		0.5	1	2	$f_r = $ ___	5	10	20	50	100
$R = 1\ \text{k}\Omega$	Output Voltage (V)									
	Gain $= V_{out}/V_{in}$									
	20 log V_{out}/V_{in} (dB)									
$R = 2\ \text{k}\Omega$	Output Voltage (V)									
	Gain $= V_{out}/V_{in}$									
	20 log V_{out}/V_{in} (dB)									

TABLE 49-3

$C = 0.001\ \mu\text{F}$		Frequency Response: $V_{in} =$								
Frequency (kHz)		0.5	1	2	5	10	$f_r = $ ___	20	50	100
$R = 1\ \text{k}\Omega$	Output Voltage (V)									
	Gain $= V_{out}/V_{in}$									
	20 log V_{out}/V_{in} (dB)									
$R = 2\ \text{k}\Omega$	Output Voltage (V)									
	Gain $= V_{out}/V_{in}$									
	20 log V_{out}/V_{in} (dB)									

NOTES

QUESTIONS FOR EXPERIMENT 49

( )

1. At the resonance frequency, if the ac coil resistance were equal to the value of R, the voltage gain ratio would be equal to
   (a) 1   **(b)** 0.707   **(c)** 0.5   **(d)** 2

2. For a given coil resistance, the maximum value of V_{out} is
   (a) less for a larger value of R
   (b) less for a smaller value of R

( )

   (c) independent of R

3. In a series-resonant band-pass filter, the bandwidth is a function of
   (a) L and C only   **(b)** C and R only   **(c)** R only

( )

   (d) R and L only

4. At resonance, if $V_{in} = 1$ V and $V_{out} = 0.55$ V across $R = 2$ kΩ, then the ac resistance of the coil is

( )

   (a) 275 Ω **(b)** 450 Ω **(c)** 225 Ω **(d)** 1636 Ω

5. Show with the aid of formulas how the resonant frequency and bandwidth of such a filter can be changed independently of one another.

6. Explain why, in the actual circuit, the bandwidth will nearly always be larger than that predicted in Table 49-1.

50

BAND-STOP FILTERS

REFERENCE READING

Principles of Electric Circuits: Section 18–4.

RELATED PROBLEMS FROM *PRINCIPLES OF ELECTRIC CIRCUITS*

Chapter 18, Problems 24 through 26.

OBJECTIVE

To determine the frequency characteristics of parallel-resonant band-stop filters.

EQUIPMENT

Audio signal generator
Oscilloscope and $10\times$ probe
High-impedance millivoltmeter
Capacitors: 0.01 μF, 0.001 μF
Inductor: 100–200 mH radio frequency coil
Resistors (±5%): 100 Ω
     1 kΩ
     2 kΩ

BACKGROUND

The band-stop filter performs the inverse operation of the band-pass filter studied in experiment 49. It passes low and high frequencies and heavily attenuates a narrow range of frequencies around a central value. Once again, resonant circuits can be used in producing such a response; on this occasion, we will use a parallel resonant circuit to demonstrate the response. With reference to Figure 50-1, the input voltage V_{in} divides between the "tank circuit" and R. When the RLC parallel network resonates, its impedance is a maximum, and therefore the output voltage will

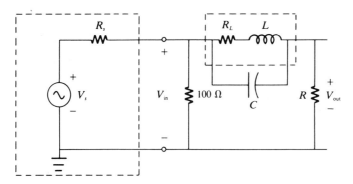

FIGURE 50-1

be at its smallest. At low frequencies, providing the ac coil resistance is small, the output voltage will very nearly equal the input voltage. This will occur similarly at high frequencies because of C's low reactance. From the references, the true resonant frequency is affected by coil resistance, as is its impedance at resonance. This will be important in determining the exact voltage gain at resonance (see Questions).

PROCEDURE

If you are using high-impedance meters for the voltage measurements in this experiment, assume amplitude values are rms. If an oscilloscope is being used for measurement, assume amplitude values are peak. Be sure to compensate any $10\times$ probes that you are going to use in this experiment; otherwise your measurements will be invalid.

1. Measure the actual values of the inductor and each of the capacitors, and, ignoring coil resistance, use these to calculate the expected resonant frequency f_r for each row of Table 50-1. Enter the measured values of R and C in the appropriate blank spaces below the corresponding nominal values. Enter the value of L under the L heading. When coil resistance is small, the practical parallel resonant frequency is the same as the series resonant frequency $1/(2\pi\sqrt{LC})$.
2. Construct the circuit, with values in the first row of Table 50-1, and, with the oscilloscope in the dual-trace mode, monitor the filter's input voltage (same as the generator's output voltage) on channel 1 and the output voltage (across the resistor) on channel 2.
3. At a frequency of 500 Hz, adjust the filter's input voltage for an amplitude of 2 V. (If you are working with voltmeters, treat this as an rms amplitude; otherwise, if exclusively using the oscilloscope, treat it as a peak amplitude.)
4. At this frequency, the output voltage of the filter should be measurable, and it ought to lag the input voltage by some angle less than 90°.
5. Measure, and record in Table 50-2, the output voltage of the filter at this frequency.
6. Locate and record, under the Measured heading in Table 50-1, the resonant frequency of the filter. You can do this by looking for the frequency at which the output voltage reaches a *minimum*, or use the fact that the phase difference between the input and output voltages is approximately 0° at this frequency. Record also V_{out} at this frequency on Table 50-2.
7. Now adjust the frequency of the generator for 1 kHz, and then, for each of the remaining frequencies in Table 50-2, measure and record the magnitude of the

output voltage relative to the input, while **maintaining V_{in} constant at 2 V.**
Watch this carefully because the generator's output voltage will tend to increase as you approach the circuit's resonant frequency. As the frequency increases, the resistor voltage V_R should at first decrease and then increase as you pass over the parallel resonant frequency.

8. Complete the Gain (V_{out}/V_{in}) and dB rows of Table 50-2.
9. Repeat the frequency response for each of the remaining pairs of resistor and capacitor values in Table 50-1, recording pertinent data in this table and in Tables 50-2 and 50-3.
10. Graph paper is provided to plot the voltage gain either directly or in dB versus (log) frequency for your measured data. See your instructor for the plots actually required.

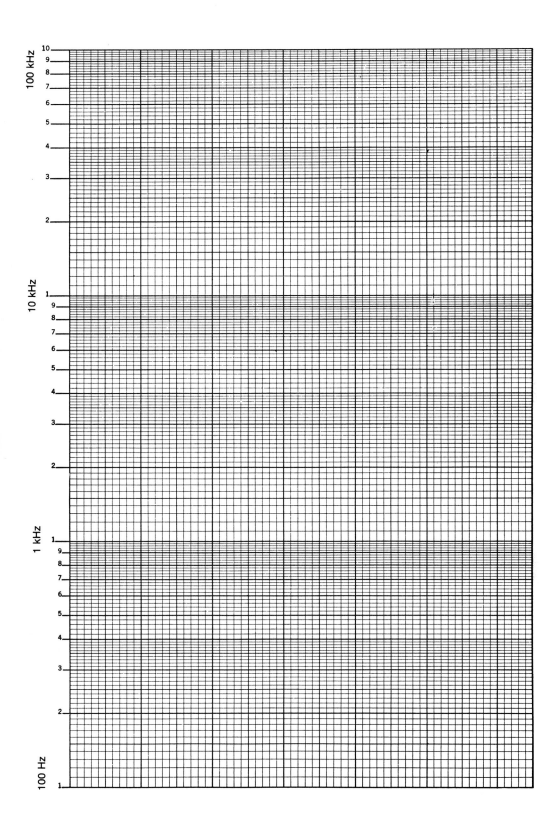

100 kHz

10 kHz

1 kHz

100 Hz

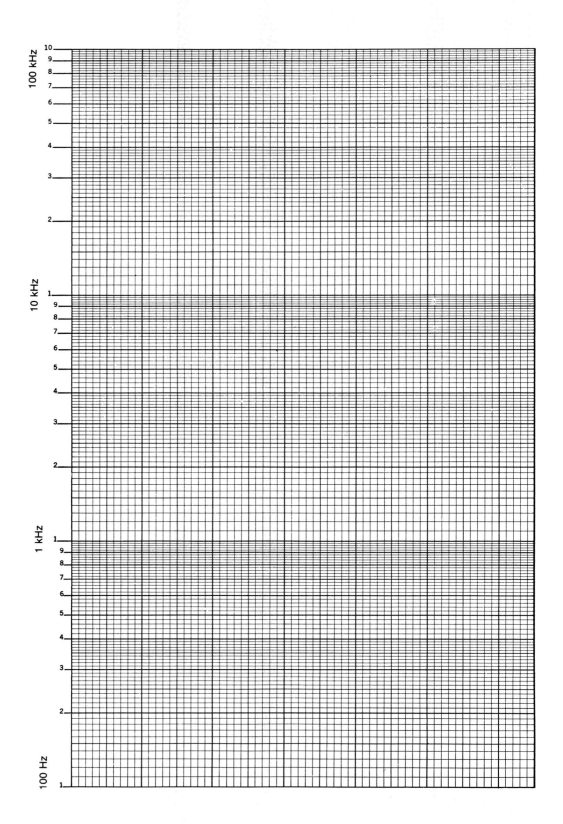

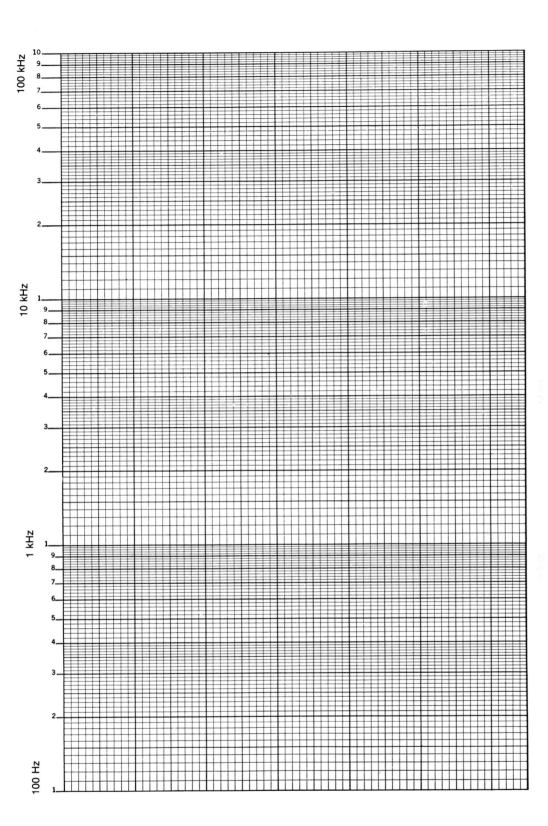

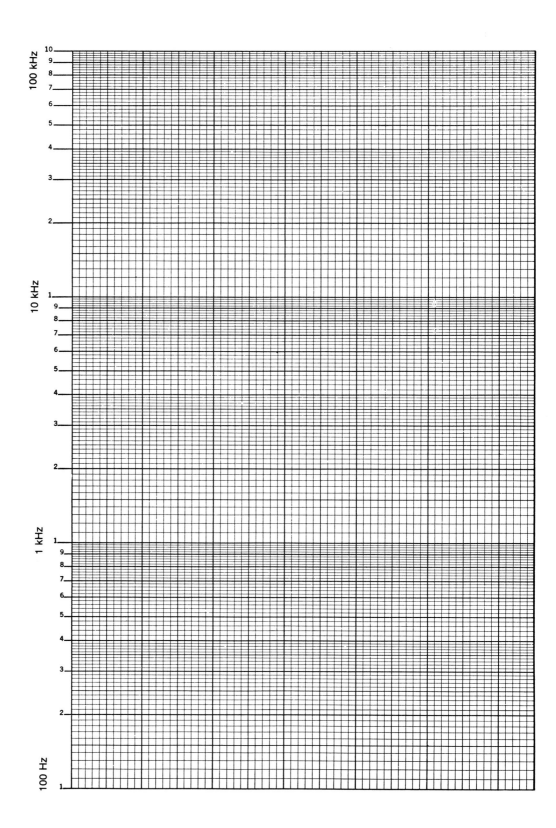

Name _____ Date _____

DATA FOR EXPERIMENT 50

TABLE 50-1

		Component Values		Resonant Frequency	
	L	C	R	Calculated	Measured
1		0.01 μF	1 kΩ		
2		0.01 μF	2 kΩ		
3		0.001 μF	1 kΩ		
4		0.001 μF	2 kΩ		

TABLE 50-2

$C = 0.01$ μF		Frequency Response V_{in} =								
Frequency (kHz)		0.5	1	2	f_r	5	10	20	50	100
$R = 1$ kΩ	V_{out}									
	Gain = V_{out}/V_{in}									
	20 log V_{out}/V_{in} (dB)									
$R = 2$ kΩ	V_{out}									
	Gain = V_{out}/V_{in}									
	20 log V_{out}/V_{in} (dB)									

TABLE 50-3

$C = 0.001\ \mu\text{F}$		Frequency Response V_{in} =								
Frequency (kHz)		0.5	1	2	5	10	f_r	20	50	100
$R = 1\ \text{k}\Omega$	V_{out}									
	Gain = V_{out}/V_{in}									
	20 log V_{out}/V_{in} (dB)									
$R = 2\ \text{k}\Omega$	V_{out}									
	Gain = V_{out}/V_{in}									
	20 log V_{out}/V_{in} (dB)									

NOTES

QUESTIONS FOR EXPERIMENT 50

1. In this band-stop filter, the output voltage
   (a) lags the input voltage
   (b) leads the input voltage
( ) (c) lags or leads the input voltage depending on frequency
2. Ideally (assuming zero coil resistance), the output voltage of such a filter at the resonant frequency ought to be
   (a) equal to the input voltage
   (b) equal to one-half the filter's input voltage
( ) (c) equal to zero
3. When coil resistance is taken into account, then the actual resonant frequency of the filter is
   (a) less than that predicted by $1/(2\pi\sqrt{LC})$
   (b) greater than that predicted by $1/(2\pi\sqrt{LC})$
( ) (c) equal to that predicted by $1/(2\pi\sqrt{LC})$
4. If $L = 150$ mH, $C = 0.01$ μF, $R = 1$ kΩ, and coil resistance were also 1 kΩ, then the filter's resonant frequency (see text reference) would be about
   (a) 4.109 kHz   (b) 3.97 kHz   (c) 24.94 kHz
( ) (d) 25.82 kHz
5. With regard to this filter circuit, explain in your own words how R can be used to control the filter's bandwidth and Q factor.

6. If the output voltage of the filter were to be taken from the tank (parallel LC) circuit instead of the resistor, what kind of filter response would we have?

51

SUPERPOSITION USING ac AND dc

REFERENCE READING

Principles of Electric Circuits: Sections 11–8, 19–1.

RELATED PROBLEMS FROM *PRINCIPLES OF ELECTRIC CIRCUITS*

Chapter 11, Problems 29 through 32. Chapter 19, Problems 1 through 5.

OBJECTIVE

To be able to predict and verify voltages in circuits containing dc and ac sources.

EQUIPMENT

Function generator with dc offset
Oscilloscope and $10 \times$ probe
DMM
dc power supply 0–10 V (if dc offset not available on function generator)
Capacitor: 0.1 μF
Resistors ($\pm 5\%$): 20 kΩ (two)

BACKGROUND

We often encounter circuits in which voltages and currents are made up of both dc and ac components; that is, they are energized from dc and ac sources simultaneously. Actually, the dc offset control on a function generator provides us with a means of controlling the dc component of an ac signal. Signals such as these are sometimes referred to as "ac riding on top of dc," and when you look at such waveforms with the oscilloscope, you can see why the description is an appropriate one.

To analyze reactive circuits when more than one source is involved, the technique of superposition is employed. As in dc circuits, the response due to each of the

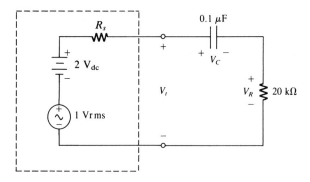

FIGURE 51-1

sources acting alone is found, and then the results are added to get the complete response when all sources are acting simultaneously. In this method, voltage sources, which are considered inactive, must be replaced by short circuits.

Two such circuits will be looked at in this experiment. The first, a *coupling* circuit, is shown in Figure 51-1. Notice that the circuit contains one resistor, a capacitor, an ac source (the function generator), and a dc source (the power supply). The dc source can actually be the dc offset control on the function generator if it has one. However, in this discussion, we will assume that it is a separate supply. The purpose of the capacitor is to allow the ac part of the signal supplied by the function generator to be developed across the 20 kΩ resistor, while at the same time *blocking any dc current* that might otherwise flow *out of the power supply*. The capacitor then, *couples the ac and blocks the dc*. If the reactance of the capacitor is small enough at the frequency of the *ac source,* it can, to an approximation, be considered a *short circuit;* to any *dc* coming from the power supply, it is an *open circuit.* The voltages and currents in the circuit can be calculated using superposition and adding the separate dc and ac effects to get the total effect. Incidentally, this circuit is a very common application of the capacitor, and you will see it over and over when you reach a more advanced stage of study.

The second application, a *bypassing* circuit, is shown in Figure 51-2. Notice that this circuit contains two resistors, a capacitor, an ac source, and a dc source. The purpose of the capacitor is to force almost all of the ac signal voltage supplied by the function generator to be developed across the upper 20 kΩ resistor and none across the lower 20 kΩ resistor, while at the same time allowing the dc portion of the generator voltage to be developed equally across the 20 kΩ resistors. The capacitor then *bypasses the ac and blocks the dc*. If the reactance of the capacitor is small

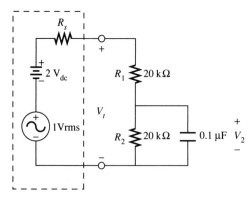

FIGURE 51-2

enough at the frequency of the *ac source,* it can, to an approximation, be considered a *short circuit;* to any *dc* coming from the dc portion of the input signal, it is an *open circuit.* Once again, the voltages and currents in the circuit can be calculated using superposition and adding the separate dc and ac effects to get the total effect.

Finally this is a good opportunity to further explore use of the oscilloscope input coupling switch in both its dc and ac modes.

PROCEDURE

Find out whether the DMMs in your laboratory are good to 10 kHz on their ac ranges. If not, then use the oscilloscope for all of the ac measurements in this activity. Be sure to compensate any $10\times$ probes that you might use in this experiment; otherwise your measurements will be invalid.

Part A: A Coupling Circuit

1. Use the superposition principle to calculate the dc component of the voltages across both R and C in the circuit of Figure 51-1. Assume that the dc source in the dotted box has a value of 2 V dc, and treat the ac source within the box as a short circuit. The source resistance R_s appears in the circuit, but because the capacitor is an open circuit to dc, you will not need to know the value of R_s to do any of the calculations. Record as *calculated* dc data in Table 51-1.
2. Use the superposition principle again to calculate the *peak values* of the ac portion of the voltages across R and C for a voltage of 0.7 V rms (1 V peak). Assume that the reactance of C is negligibly small at 10 kHz and treat it as a short. The source resistance R_s appears in the circuit, but, since it is generally much less than 20 kΩ, simply ignore it (assume $R_s = 0$ Ω) for these calculations. Record as *calculated* ac data in Table 51-1.
3. Now set the function generator for a sine-wave output at 10 kHz and rms amplitude 0.7 V. If it accurately responds at frequencies as high as 10 kHz, use a DMM for this; otherwise use the oscilloscope to set up the ac voltage for a p–p amplitude of 2 V.
4. Now set up the dc component of the total signal in one of the two following ways:
   (a) If you have a dc offset control on the function generator, position a DMM set to **dc VOLTS** across the terminals, and adjust the dc offset until you read 2 V dc.
   (b) If no dc offset control is provided, then connect a regulated dc power supply in series with the generator so that the *negative* terminal of the supply connects to the *high* (ungrounded) side of the function generator, and the *positive* side of the supply connects to the *capacitor.* Make sure that the negative side of the dc supply is *not* grounded. Connect the DMM across the terminals of the power supply and adjust it for 2 V dc.
5. Remove the DMM, and connect the oscilloscope to view the combined dc and ac components of the signal. With input coupling on **DC,** connect channel 1 to the high side of the generator. You should see a sine wave with a p–p value of about 2 V "sitting on" a dc level of +2 V. (Why?)
6. Switch the oscilloscope to ac coupling and notice the shift in the display. How is the shift related to the components of the signal?
7. Using the DMM, measure the dc voltage across the capacitor *and* the resistor. Record in Table 51-1 in the Measured Data row.
8. Measure the (p–p) ac component of the resistor voltage, using the oscilloscope, and record in Table 51-1 in the Measured Data row. Should it matter whether or not you select dc or ac coupling for this measurement? If in doubt, try both

settings. The only thing left to measure for this table is the ac component of capacitor voltage.

9. Switch the positions of the resistor and capacitor so that the capacitor is now grounded; then connect the oscilloscope across the capacitor. With input coupling on **DC,** you should see a steady level of 2 V, and if you look carefully, you might see some small ripples (sine-wave variation) on the top. If you increase the vertical sensitivity **(VOLTS/DIV)** setting so as to get a better look at this small variation, the 2 V dc level will force the trace off the screen. Try this.

10. Here is the solution to this problem. Switch the input coupling for channel 1 to **AC,** and as you switch to a lower **VOLTS/DIV** setting, the ac component of the capacitor voltage should become visible. Measure the peak-to-peak value of this voltage, and record in Table 51-1 in the Measured Data row.

11. Repeat steps 3 through 10 at a frequency of 1 kHz and record all data in Table 51-2. Why is the ac component of the capacitor voltage larger at 1 kHz than at 10 kHz?

Part B: A Bypassing Circuit

1. Use the superposition principle to calculate the dc component of the voltages across both R_1 and R_2 in the circuit in Figure 51-2. Note that V_2 is also the capacitor voltage because they are in parallel. As in part A, treat the ac source together with R_s (negligible) as a short circuit. Assume that the dc source in the dashed box has a value of 2 V dc. Record as *calculated* dc data in Table 51-3.

2. Use the superposition principle again to calculate the peak-to-peak values of the ac portion of the voltages across R_1 and R_2. Assume that the ac source in the dashed box has a value of 0.7 V rms (2 V p–p) and treat the dc source within the box as a short circuit. Again, assume $R_s = 0\ \Omega$ and $X_C = 0\ \Omega$ for these calculations. Record as *calculated* ac data in Table 51-3.

3. Now set the function generator for a sine-wave output at 10 kHz and rms amplitude 0.7 V (2 V p–p). If it accurately responds at frequencies as high as 10 kHz, use a DMM for this; otherwise use the oscilloscope to set up the ac voltage for a p–p amplitude of 2 V.

4. Now set up the 2 V dc component of the signal, using either the offset control or a separate power supply.

5. Connect the oscilloscope to view the combined dc and ac components of the signal. With input coupling on dc, connect channel 1 to the ungrounded side of the generator. You should see a sine wave with a p–p value of about 2 V "sitting on" a dc level of +2 V.

6. Using the DMM, measure the dc voltages across R_1 and R_2. Record in Table 51-3 in the Measured Data row.

7. Measure the ac component of voltage V_2, using the oscilloscope, and record in Table 51-3 in the Measured Data row. This value will be very small because of the low impedance of the bypass capacitor. Nevertheless, it is measurable. To see the ac component, you will need to set the input coupling switch to **AC** so as to filter out the dc level and then switch to a sensitive **VOLTS/DIV** setting.

8. Because both resistors have the same value, in order to measure the ac component of V_1, you can simply switch the capacitor so that it is parallel with R_1, and measure the voltage V_2 instead. *The voltage* V_2 *will now be the same as* V_1 *was when it had no bypass capacitor.* With input coupling on **DC,** you should see the full ac voltage coming from the generator sitting on top of the dc level. Record the ac component of this voltage in Table 51-3.

9. Repeat steps 3 through 8 at a frequency of 1 kHz and record all data in Table 51-4. Why is the ac component of the capacitor voltage larger at 1 kHz than at 10 kHz?

DATA FOR EXPERIMENT 51

TABLE 51-1

	V_C		V_R	
	dc	ac	dc	ac
Calculated Data				
Measured Data				

TABLE 51-2

	V_C		V_R	
	dc	ac	dc	ac
Calculated Data				
Measured Data				

TABLE 51-3

	V_{R_1}		V_{R_2}	
	dc	ac	dc	ac
Calculated Data				
Measured Data				

TABLE 51-4

	V_{R_1}		V_{R_2}	
	dc	ac	dc	ac
Calculated Data				
Measured Data				

NOTES

QUESTIONS FOR EXPERIMENT 51

(   ) **1.** The capacitive reactance of the capacitor at 10 kHz is closer to
                      (a) 150 Ω **(b)** 1 kΩ **(c)** 10 kΩ **(d)** 10 Ω

                      2. The purpose of the capacitor in Figure 51-1 is to
(a) block the dc voltage from R
(b) block the ac voltage from R
(c) **a** and **b** are both correct
(d) allow the ac voltage to appear across R

(   ) **(e)** **a** and **d** are both correct

                      3. With reference to Figure 51-2, which of the following statements is correct?
(a) The dc voltage across R_1 is 0 V.
(b) The ac voltage across R_2 is approximately 0 V.
(c) The dc current through C is 0 A.
(d) The ac current through R_2 is approximately 0 A.

(   ) **(e)** **b, c,** and **d** are correct.

                      4. The best form of coupling when viewing a *small ac signal* superimposed upon a relatively large dc level is

(   )   **(a)** ac   **(b)** dc   **(c)** both are equally good

                      5. The resistor in Figure 51-1 is said to be ac coupled to the source by the capacitor. In the light of your data, attempt to explain this statement.

                      6. The resistor R_2 in Figure 51-2 is said to be bypassed by the capacitor. In the light of your data, attempt to explain this statement.

52

THE SERIES *RC* CIRCUIT PULSE RESPONSE

REFERENCE READING

Principles of Electric Circuits: Sections 20–1 through 20–5.

RELATED PROBLEMS FROM *PRINCIPLES OF ELECTRIC CIRCUITS*

Chapter 20, Problems 1 through 17, 25, and 26.

OBJECTIVE

To examine the step and pulse response of a series *RC* circuit.

EQUIPMENT

Square-wave generator
Oscilloscope and two $10\times$ probes
Capacitor ($\pm 10\%$): 0.001 μF
Resistor ($\pm 5\%$): 20 kΩ

BACKGROUND

The *RC* circuit in Figure 52-1 has been looked at in terms of both its low-pass and high-pass filtering aspects. The circuit can also be used to change the shape of pulse waveforms. In this experiment, we will observe the response of an *RC* circuit to a repetitive voltage pulse. The pulse will be square and therefore have a 50 percent duty cycle. The voltages across the capacitor and resistor in this case will depend on the relative magnitudes of the time constant and the pulse width. In the first part of the experiment, the frequency will be such that the capacitor has ample time

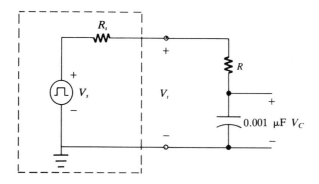

FIGURE 52-1

to charge and discharge for each half-period of the input signal. You will be able to confirm the equations for the rising and falling portions of the capacitor waveform.

In the second part of the experiment, the frequency will be increased so that the capacitor does not have sufficient time to reach a steady-state (dc) value. In this case, an ac steady-state condition is reached; the capacitor voltage oscillates between two values equally spaced around the average value of the input waveform. As the frequency is increased still further, the capacitor voltage approaches an almost constant value which, again, is equal to the average value of the input waveform.

PROCEDURE

Be sure to compensate any $10\times$ probes that you are going to use in this experiment; otherwise your measurements will be invalid.

Part A: Pulse Response When Capacitor Has Ample Time to Fully Charge/Discharge

1. For the RC combination in Table 52-1, calculate and record the time constant and the approximate total charging time.
2. Assuming a square-wave input to the circuit in Figure 52-1, calculate and record the frequency f_{max} that will just allow sufficient time for the capacitor to fully charge (5τ) and then fully discharge (another 5τ) for each cycle of the input waveform.
3. Construct the circuit in Figure 52-1. Set up simultaneous Channel 1 and 2 ground reference traces on the lowest horizontal line of the graticule.
4. Set the Channel 1 and 2 vertical sensitivity controls to 0.5 **VOLTS/DIV.** With the oscilloscope channel 1 connected to monitor the generator's terminal voltage, and channel 2 to monitor the capacitor voltage, adjust the output for a 1 kHz square wave of p–p amplitude 4 V and baseline 0 V. You will need to use the offset control to insert a dc level of 2.0 V into this signal, or else use a dc power supply as in the past. Set up the oscilloscope for dual-trace operation, and trigger from channel 1.
5. Observe both the terminal voltage and capacitor voltage simultaneously, and note that the capacitor has ample time to charge and discharge at this frequency.
6. Gradually increase the frequency toward 5 kHz, and watch the capacitor voltage. At 5 kHz, the capacitor has just enough time to charge to $+4$ V and back down to 0 V within one period of the waveform. This frequency should be the one you have recorded as f_{max} in Table 52-1.

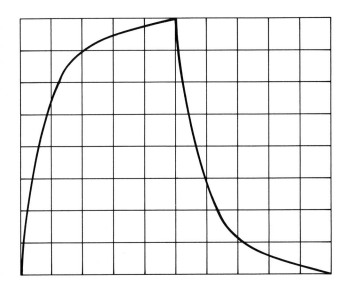

FIGURE 52-2

7. Set up the oscilloscope to view only the capacitor voltage (channel 2). With the horizontal sensitivity at 20 **μs/DIV,** use the horizontal **POSITION** control together with the trigger **SLOPE** and **LEVEL** controls to display exactly one full charge-discharge cycle on the screen, as shown in Figure 52-2. You may need to fine-tune the frequency a little to get *exactly* one cycle across the width of the screen.

8. Using the equation for a rising exponential, calculate the expected voltage values for V_C at $t = 1, 2, 3, 4$, and 5 time constants into the charging period and record these values as calculated data in Table 52-2.

9. Set the horizontal sensitivity to 10 **μs/DIV.** At this setting, one time constant is represented by approximately *two* major divisions across the screen ("approximately" because this assumes that the resistor and capacitor are equal to their nominal values). Use the oscilloscope to determine, as best as you can, the actual capacitor voltages at 1, 2, 3, 4, and 5 time constants into the charging period, and record these values as measured data in Table 52-2.

10. Using the equation for a falling exponential, calculate the expected voltage values for V_C at $t = 1, 2, 3, 4$, and 5 time constants into the discharging period and record these values as calculated data in Table 52-3.

11. With horizontal sensitivity at 10 **μs/DIV,** by triggering on a negative **SLOPE** and adjusting the **LEVEL** control, bring the *discharge* portion of the waveform into view. Use the oscilloscope to determine, as best as you can, the actual capacitor voltages at 1, 2, 3, 4, and 5 time constants into the discharging period, and record these values as measured data in Table 52-3.

12. Restore the horizontal sensitivity to 20 **μs/DIV.** Sketch one full cycle of the capacitor voltage in the space provided in Figure 52-3, then exchange the resistor and capacitor so that the resistor is grounded (or else use the **ADD/ INVERT** feature on the oscilloscope). Use channel 2 to view the resistor voltage, and sketch one full cycle of the waveform in Figure 52-3. Note that you will have to change the vertical sensitivity to 1 **VOLTS/DIV** to see the complete resistor waveform on the screen. Label the voltage and time axes in your sketches. You will need this information in the questions at the end of the experiment.

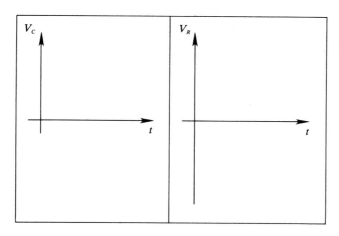

FIGURE 52-3

Part B: Pulse Response When Capacitor Has Insufficient Time to Fully Charge/Discharge

1. Restore the capacitor to the grounded position, and adjust the oscilloscope controls so that five complete cycles of the capacitor voltage are displayed across the screen. It may help at this point to temporarily remove the channel 1 display if you wish. Record the peak-to-peak capacitor voltage in Table 52-4.

2. Increase the frequency to 25 kHz and observe the capacitor voltage. Its peak-to-peak value should have decreased. The capacitor no longer has sufficient time to fully charge and discharge each cycle, and the steady-state voltage therefore oscillates on either side of the 2 V dc component of the waveform. Record the peak-to-peak value in Table 52-4.

3. Continue to increase the frequency through the remaining values in the table, recording the peak-to-peak voltage across the capacitor in each case. At high frequencies, this voltage becomes very very small. Note that if you want to measure the peak-to-peak value accurately you will need to select **AC** coupling and move the ground reference to the center of the screen.

4. Continue raising the frequency, and note what happens to the capacitor voltage.

5. Exchange the positions of the resistor and capacitor, and observe the resistor voltage at each of the frequencies in Table 52-4. The waveform should become more square as the frequency increases. At very high frequencies, what is the only difference between the input and resistor voltages? (*Hint:* Flip input coupling from **DC** to **AC** and back when looking at this voltage.)

Name _____ Date _____

DATA FOR EXPERIMENT 52

TABLE 52-1

	R	C	τ	5τ	f_{max}
1	20 kΩ	0.001 μF			

TABLE 52-2 *Charging curve*

Number of Time Constants	Calculated Voltage	Measured Voltage
1		
2		
3		
4		
5		

TABLE 52-3 *Discharging curve*

Number of Time Constants	Calculated Voltage	Measured Voltage
1		
2		
3		
4		
5		

TABLE 52-4

Frequency (kHz)	$V_{\text{p-p}}$ (V)
5	
10	
25	
50	
100	
200	

NOTES

QUESTIONS FOR EXPERIMENT 52

1. In a circuit such as that in Figure 52-1, suppose $R = 50 \text{ k}\Omega$ and $C = 0.01 \text{ μF}$. The value of f_{max} is

( )   (a) 4 Hz   (b) 400 kHz   (c) 400 Hz   (d) 200 Hz

2. In this experiment the source resistance was ignored. It actually
   (a) increases the effective time constant
   (b) decreases the effective time constant
( )   (c) makes no contribution to the time constant

3. When you raised the frequency to very high values in step 4 of part B of the procedure, the voltage V_c
   (a) became closer to 0 V
   (b) became closer to 4 V
( )   (c) became closer to 2 V

4. For all values of frequency less than f_{max}, the capacitor will have enough time to fully charge and discharge in each cycle.
( )   (a) True   (b) False

5. Explain, with the aid of sketches, why the peak-to-peak resistor voltage is twice that of the input square wave.

6. Explain why the charge/discharge time is independent of the input signal amplitude.

53

BALANCED THREE-PHASE SYSTEMS

REFERENCE READING

Principles of Electric Circuits: Sections 21–3, 21–4.

RELATED PROBLEMS FROM *PRINCIPLES OF ELECTRIC CIRCUITS*

Chapter 21, Problems 6 through 16.

OBJECTIVE

To observe the line-phase relationships of the currents and voltages in balanced three-phase systems.

EQUIPMENT

Three-phase 120 V/208 V power system
ac voltmeter and ammeters
Resistors (½ W ±5%): three 68 kΩ

BACKGROUND

Three-phase electrical systems have several advantages over single-phase types, as discussed in the references. The three generators may be connected end-to-end, forming a delta- (Δ-) connected system. Alternatively, they may be connected as a wye (Y), where three ends have a common reference, and the other ends each have a potential with respect to this point. The phases might be referred to with letters (A, B, C), numbers (1, 2, 3), or colors (red, yellow, blue).

In a delta-connected system, the line currents are different from the phase currents; the line and phase voltages are, however, the same. In a wye-connected system, the opposite is true; that is, the line and phase voltages are different, but the line and phase currents are equal.

When the load resistances are equal, the system is said to be "balanced." This is the case we are going to investigate in this experiment.

PROCEDURE

NOTE: **THE VOLTAGES IN A THREE-PHASE SYSTEM ARE POTENTIALLY LETHAL. BE SURE TO HAVE THE INSTRUCTOR CHECK YOUR CONNECTIONS *BEFORE* YOU SWITCH ON THE POWER.**

Part A: Balanced Δ System

1. Enter the line/phase voltages in Table 53-1.
2. Calculate and record the phase and line currents for the given loads in Figure 53-1.
3. Connect the circuit in Figure 53-1. Be sure to keep your hands away from exposed connections once power is applied. Measure and record the line currents, phase currents, and line/phase voltages. **WHEN MOVING THE AMMETER POSITION, SWITCH OFF THE POWER.**

Part B: Balanced Y System

1. Enter the line and phase voltages in Table 53-2.
2. Calculate and record the line/phase currents for the given loads in Figure 53-2.
3. Connect the circuit in Figure 53-2. Once more, be sure to keep your hands away from exposed connectors when power is connected. Measure and record the line voltages, phase voltages, and line/phase currents.

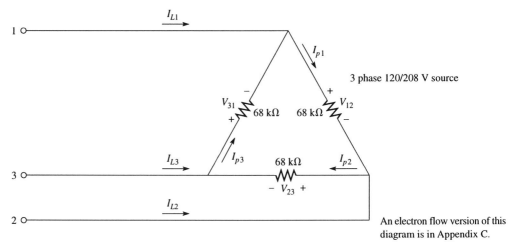

FIGURE 53-1

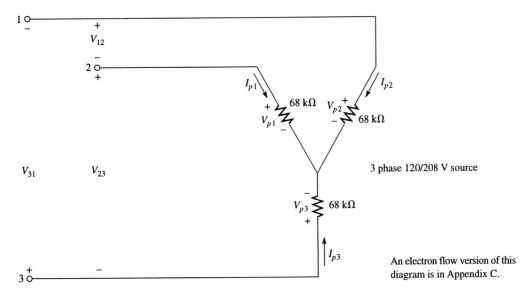

1 $\circ$ —
$+$
V_{12}
$-$

2 $\circ$ —
$-$
$+$

I_{p1}
$+$ 68 kΩ V_{p2} $+$
V_{p1} $-$ 68 kΩ
$-$

I_{p2}

V_{31} $\qquad$ V_{23}

3 phase 120/208 V source

$-$
V_{p3} 68 kΩ
$+$

$+$ $\qquad$ $-$
3 $\circ$ —

I_{p3}

An electron flow version of this
diagram is in Appendix C.

FIGURE 53-2

Name _____ Date _____

DATA FOR EXPERIMENT 53

TABLE 53-1

	V_{12}	V_{23}	V_{31}	I_{p1}	I_{p2}	I_{p3}	I_{L1}	I_{L2}	I_{L3}
Calculated Data									
Measured Data									

TABLE 53-2

	V_{p1}	V_{p2}	V_{p3}	V_{12}	V_{23}	V_{31}	I_{L1}	I_{L2}	I_{L3}
Calculated Data									
Measured Data									

NOTES

QUESTIONS FOR EXPERIMENT 53

1. In a balanced Δ system
   (a) the line and phase voltages are equal
   ( )   (b) the line and phase currents are equal

2. In a balanced Y system
   (a) the line and phase voltages are equal
   ( )   (b) the line and phase currents are equal

3. Which of the statements is true for a balanced system?
   (a) A Δ system has 3 or 4 wires.
   (b) A Y system has 3 or 4 wires.
   (c) A Y system needs only 3 wires.
   ( )   (d) b and c are both true.

4. The total power in the system in Figure 53-1 is about
   ( )   (a) 212 mW   (b) 635 mW   (c) 700 mW   (d) 1.91 W

5. Explain in your own words why the neutral wire can be removed in a balanced Y system.

( )

6. Explain in your own words why there is a saving in copper when using a three-phase system when compared with three single-phase systems.

APPENDICES

PARTS AND INSTRUMENTS FOR THE EXPERIMENTS

RESISTORS

One each (all ± 5%):

22 Ω	820 Ω	2.7 kΩ	6.8 kΩ	30 kΩ
33 Ω	1.1 kΩ	3.3 kΩ	7.5 kΩ	33 kΩ
47 Ω	1.2 kΩ	3.9 kΩ	8.2 kΩ	47 kΩ
68 Ω	1.5 kΩ	4.3 kΩ	12 kΩ	62 kΩ
100 Ω	1.6 kΩ	4.7 kΩ	15 kΩ	100 kΩ
200 Ω	1.8 kΩ	5.1 kΩ	18 kΩ	330 kΩ
470 Ω	2.2 kΩ	5.6 kΩ	22 kΩ	1 MΩ
620 Ω	2.4 kΩ	6.2 kΩ	24 kΩ	10 MΩ

Two each (all ½ W, ±5%): 10 kΩ, 16 kΩ, 20 kΩ, 200 kΩ
Three each (all ½ W, ±5%): 1 kΩ, 2 kΩ, 68 kΩ
Five each (all ½ W, ±5%): 3 kΩ
One (¼ W): 100 Ω, 200 Ω
One (2 W): 120 Ω
Three 100-Ω precision

POTENTIOMETERS

One each: 1 kΩ, 10 kΩ
At least one ten-turn

CAPACITORS (FIXED)

One each (at least 10 V rated):
 0.001 μF, 0.01 μF, 0.1 μF,
 1.0 μF, 0.22 μF, 0.47 μF,
 4.7 μF, 500 μF

INDUCTORS (FIXED)

One coil in the range 100–200 mH with $Q \geqslant 10$, over 5–12 kHz
One coil 1–5 H (low ac resistance, <500 Ω at 1 kHz)

LAMPS

No. 47 or equivalent
Neon

TRANSFORMERS

One filament-type 115–120 V/12.6 V/60 Hz
One audio impedance-matching transformer (any kind; instructor must choose
source and load resistors accordingly); e.g., universal output transformer (TRIAD
S-51X or equivalent)

FUSES

A selection of blown and intact fuses

BATTERIES

Four 1½-V dry cells

SWITCHES

NOPB, NCPB, SPST, SPDT

INSTRUMENTS

One each: VOM, DMM, audio signal generator
general purpose dual-trace oscilloscope
high-impedance millivoltmeter
Two dc power supplies 0–12 V (variable)
Three-phase supply

RESISTOR COLOR CODING AND STANDARD VALUES

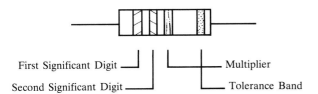

First Significant Digit —┐ ┌— Multiplier

Second Significant Digit —┘ └— Tolerance Band

FIGURE B-1 *Resistor, carbon composition type*

TABLE B-1 *Resistor color code*

Color	First Significant Digit	Second Significant Digit	Multiplier	Tolerance (percent)
Black	—	0	10^0	—
Brown	1	1	10^1	—
Red	2	2	10^2	—
Orange	3	3	10^3	—
Yellow	4	4	10^4	—
Green	5	5	10^5	—
Blue	6	6	10^6	—
Violet	7	7	10^7	—
Gray	8	8	10^8	—
White	9	9	10^9	—
Silver	—	—	0.01	10
Gold	—	—	0.1	5
No Color	—	—	—	20

TABLE B-2 *Standard values*

20%	10%	5%	20%	10%	5%
1.0	1.0	1.0	3.3	3.3	3.3
		1.1			3.6
	1.2	1.2		3.9	3.9
		1.3			4.3
1.5	1.5	1.5	4.7	4.7	4.7
		1.6			5.1
	1.8	1.8		5.6	5.6
		2.0			6.2
2.2	2.2	2.2	6.8	6.8	6.8
		2.4			7.5
	2.7	2.7		8.2	8.2
		3.0			9.1

Note: These values can be in multiples of 1x, 10x, 100x, 1000x, etc.

C

ELECTRON-FLOW DIAGRAMS

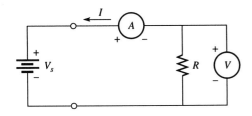

FIGURE 5-2

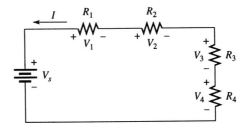

FIGURE 10-1

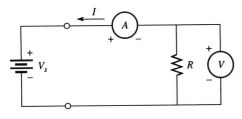

FIGURE 6-1

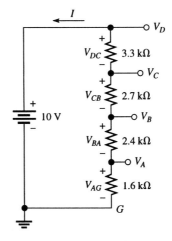

FIGURE 11-1

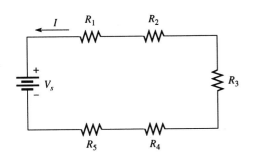

FIGURE 8-1

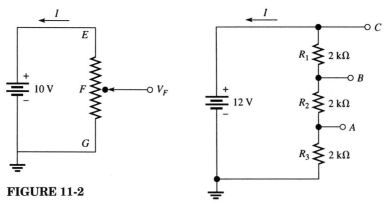

FIGURE 11-2

FIGURE 12-1

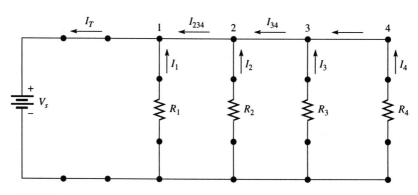

FIGURE 13-1

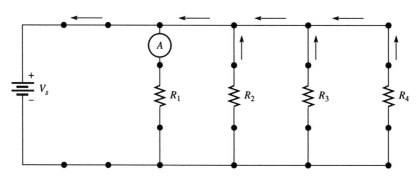

FIGURE 13-2

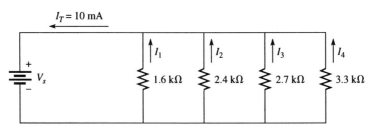

FIGURE 15-1

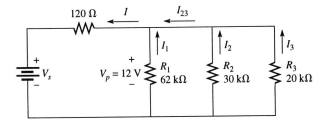

FIGURE 16-1

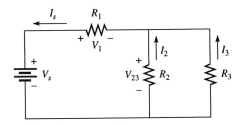

FIGURE 17-1

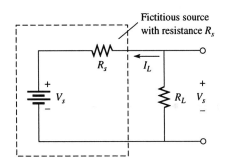

FIGURE 23-1

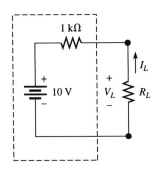

FIGURE 24-1

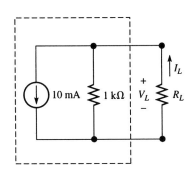

FIGURE 24-2

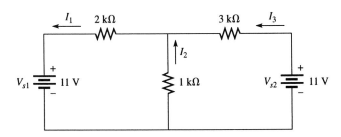

FIGURE 25-1

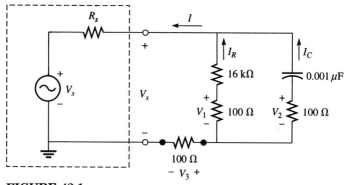

FIGURE 42-1

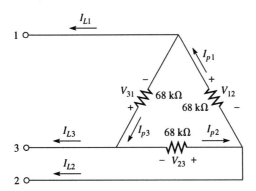

FIGURE 53-1

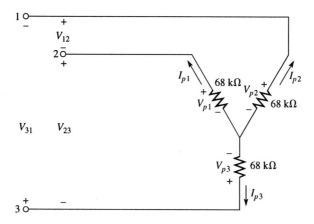

FIGURE 53-2